MATLAB Manual

Ordinary Differential Equations

MATLAB Manual

Ordinary Differential Equations

John C. Polking

Rice University

MATLAB® Curriculum Series

Prentice Hall, Englewood Cliffs, New Jersey, 07632

Production Editor: *Dirk Knoop*
Acquisitions Editor: *George Lobell*
Supplement Acquisitions Editor: *Audra Walsh*
Production Coordinator: *Alan Fischer*

Printed in the United States of America

10 9 8 7 6 5 4 3 2 1

ISBN 0-13-133944-3

Prentice-Hall International (UK) Limited, *London*
Prentice-Hall of Australia Pty. Limited, *Sydney*
Prentice-Hall Canada Inc., *Toronto*
Prentice-Hall Hispanoamericana, S.A., *Mexico*
Prentice-Hall of India Private Limited, *New Delhi*
Prentice-Hall of Japan, Inc., *Tokyo*
Simon & Schuster Asia Pte. Ltd., *Singapore*
Editora Prentice-Hall do Brasil, Ltda., *Rio de Janeiro*

Contents

Preface

Perhaps there is no elementary college level mathematics course for which computer graphics is more useful to the student than the ordinary differential equations course. This manual was originally written to enable us to institute a significant computer component in the ODE course at Rice University.

The manual is meant to be used by the student while working at the computer. The student should follow the manual, executing the commands as they occur in the text. Pains have been taken to make it as easy as possible for the student to work through the manual with a minimum of assistance. Of course, some assistance will inevitably be required.

This is not a textbook. It should not be used as the only resource for the students in a course. The manual is not written as a companion for any particular textbook. There are many textbooks which can be used with the manual, and that choice is left to the instructor. Although the manual is basically a book on how to use MATLAB* to solve ODEs, there are theorems stated, and there are theoretical results quoted in the manual. However, there are no proofs.

I have often been asked why I chose MATLAB for our students at Rice to use in this course, instead of the excellent ODE software available. I wanted our students to spend their time learning about ODEs and not about software. That meant that the software should be as easy as possible to use. Another consideration, possibly in conflict with the first, was that if our students were going to spend time learning software, then it would be desirable if that knowledge were useful in other venues after the course was over. These thoughts led me to look closely at the mathematical computer programs MATLAB, Maple, and Mathematica.

---

* MATLAB is a registered trademark of
The MathWorks, Inc.
24 Prime Park Way
Natick, MA 01760-1500
Phone: (508) 653-1415
Fax: (508) 653-2997

After spending a lot of time working with these three, I decided that MATLAB was by far the easiest to use. There were two drawbacks. First MATLAB did not, at that time, have a symbolic computing capability, but I decided that that was not as important as the ability to solve ODEs, and to graphically display the solutions. MATLAB does this better than the other two programs. Furthermore, The Symbolic Math Toolbox has since become available from The MathWorks. The second drawback was more serious. When teaching ODEs, it is highly desirable to have easy to use routines that allow students to observe the direction fields of first order ODEs, and, at a more advanced level, to observe the vector fields of planar autonomous systems. MATLAB had neither of these, but I had been using MATLAB for some time, and I knew that it would not be difficult to provide these routines for our students by writing programs in MATLAB.

So I chose MATLAB. I have been pleased with the results. By all indications, our students have been pleased as well.

Originally the programs I wrote would only run with the professional version of MATLAB version 4, but now the student edition has been upgraded, and they run on both the professional and the student versions. There is nothing in this manual that will not run on *The Student Edition of MATLAB v4* (available from Prentice Hall).

The emphasis in the manual is on ODEs. The features of MATLAB are explained only in so far as they apply to the study of ODEs. Given that as a guiding principle, the user of this manual will nevertheless learn a great deal about MATLAB in the process. There is no programming required in the use of the manual, unless you call

```
function xpr = xsqmt(t,x)

xpr = x^2 - t;
```

a program. I do not. There are some exercises that require what I do call programming, but in these cases the code is provided.

There are a large number of exercises. They range in difficulty from the routine to what should more properly be called computer projects. Not all of the questions are directly computer related. Many of them require the student to use the computer to make a conjecture, and then to verify their conjecture analytically. Where a problem requires the student to use aspects of MATLAB which are not explained in the text, these are explained as part of the problem. The textbook being used is another source of problems, especially for Chapters four and seven, where the symbolic capabilities of MATLAB are described.

The first chapter is extremely elementary. It is meant for the user with almost no experience with computers.

It is in the second chapter that we begin with ODEs. This chapter describes the MATLAB function `dfield`, which displays the direction field of a single, user provided ODE of first order. This function is very easy to use, and does not require any knowledge of how MATLAB works beyond knowing how to start it. Using `dfield`, many of the important concepts in ODEs can be explained and illustrated.

The third chapter is fundamental to almost everything that follows. It is here that the workings of MATLAB are explained. It is not necessary at this point to go very far into MATLAB. It is only necessary to understand how MATLAB works with matrices, and how to write function M-files of the complexity illustrated above. The graphical features of MATLAB are explored. Students learn this material easily.

Chapter four provides an introduction to the symbolic capabilities of MATLAB. The discussion here is limited to the solution and analysis of equations of first order. Chapter five uses MATLAB to explore numerical methods of solving ODEs. Using MATLAB, it is quite easy to explore the errors actually made when the standard one-step solution methods are used.

Chapter six is one of the most important chapters in the manual. It is here that the use of MATLAB's own ODE solvers is described, and that the student will learn how to solve essentially arbitrary systems of ODEs numerically. Learning this skill should be a goal of a modern ODE course. It is something that the student can take from the course and use in other situations.

In Chapter seven we return to MATLAB's symbolic capabilities, and develop them further. We discuss how to solve higher order equations and simple systems, and how to use the symbolic and graphical power of MATLAB to analyze the solutions.

Chapter eight discusses linear algebra. More specifically the topic is the use of MATLAB to solve linear systems of equations. Many ODE books assume that students know this important topic, an assumption which is often unwarranted. For that reason the discussion in Chapter eight is fairly complete.

In Chapter nine we apply the material in Chapter eight to the solution of linear systems of ODEs with constant coefficients. The ease with which MATLAB does matrix algebra makes it possible for students to come to grips with a wider vairety of systems than they would without MATLAB.

Planar autonomous systems are discussed in Chapter ten. A new function is introduced, called `pplane`, which is an easy to use tool which plots the direction field for a user defined system, and allows easy computation and display of solution curves. `pplane` enables the user to explore many aspects of planar systems, including finding and classifying equilibrium points, and plotting stable and unstable solution curves.

The order of the chapters is close to the order of the topics in many ODE books. However, the prerequisites for the chapters are frequently minimal. Here is a complete list of the major dependencies:

$$1 \to 2$$
$$1 \to 3 \to 4 \to 7$$
$$1 \to 3 \to 5$$
$$1 \to 3 \to 6$$
$$1 \to 3 \to 8 \to 9$$
$$1 \to 10$$

It is clear that the material can be covered in many orders. This having been said, it must be added that toward the end of many of the chapters there is more advanced material, especially in the exercises. This is particularly true in Chapter six, where, having learned how to solve ODEs in great generality, it is natural to explore that skill. It might have been more appropriate to add a Chapter eleven devoted to computer projects — perhaps next time.

Chapter ten, and the program `pplane` described there, can be used as early as indicated. However, some of the features of `pplane` will not be useful to students until they learn more about such topics as equilibrium points and stability. Certainly, understanding many of the capabilities of `pplane` requires information discussed in Chapter nine, but `pplane` can be used at a more elementary level if desired.

There is special software that is needed for the manual. This includes the functions `dfield` and `pplane` described in Chapters two and ten, respectively, together with their associated solvers, and the very simple solver routines `eul`, `rk2`, and `rk4` used in Chapter five. These routines do not come with either the professional or the student version of MATLAB. You may request the free set of functions, which is called the *ODE Book Tools*, on PC or Macintosh diskette from The MathWorks by mailing in the business reply card bound in this book. The functions are also available using anonymous ftp from two sources — from math.rice.edu, in the directory pub/ode, and from ftp.mathworks.com, in the directory pub/books/polking.

Both the Rice and The Mathworks sites will contain the latest versions of `dfield` and `pplane` that are considered to be stable. These may differ somewhat from what is described in the manual, since the software is always evolving. There should not be major differences, however. In addition the Rice site may contain versions that are still being worked on. The user should beware — these versions may not work at all.

There are many people who contributed to this project. I should start with James L. Kinsey, Dean of the Wiess School of Natural Sciences at Rice, who originally suggested this project, and who made available resources which made quick progress possible. Ken Richardson taught our new ODE course with me for the first year, and collaborated on every aspect of it. Joel Castellano assisted with the programming at one stage, and served as a volunteer "labie."

Over the past two years, I have had conversations with many mathematicians about this manual and the associated software. I would like to mention Henry Edwards, Herman Gollwitzer, Larry Shampine, Bob Devaney, Bob Williams, Al Taylor, and Beverly West. To those whom I did not mention, I offer my gratitude and my apologies. Special thanks go to Larry Shampine and Mark Reichelt, who allowed me to have an early view of their MATLAB ODE suite, which is partially described in Chapter six.

The manuscript was prepared in TEX, using a modified set of macros originally written by Jim Carlson. Michael Spivak was always ready with an answer to my many questions about TEX.

The students at Rice deserve my special gratitude. They suffered through two years of experimentation, complaining only when appropriate. They were very active participants in the preparation of the manual in this form.

John C. Polking

Houston, Texas
November 1994

MATLAB® Curriculum Series

Cleve Moler, The Mathworks, Inc.
Editor

The Student Edition of MATLAB® Student User Guide

The Student Edition of MATLAB® for Macintosh Computers

The Student Edition of MATLAB® for MS-DOS Personal
Computers with 5 1/4" Disks

The Student Edition of MATLAB® for MS-DOS Personal
Computers with 3 1/2" Disks

Engineering Problem Solving with MATLAB®, D. M. Etter

1.   Introduction to MATLAB

MATLAB is an interactive, numerical computation program. It has powerful built-in routines for enabling a very wide variety of computations. It also has easy to use graphics commands that make the visualization of results immediately available. In some installations MATLAB will also have a Symbolic Toolbox which allows MATLAB to perform symbolic calculations as well as numerical calculations. In this chapter we will describe how MATLAB handles simple numerical expressions and mathematical formulas.

MATLAB is available on almost every computer system. It's interface is similar regardless of the system being used. In this Manual we are going to assume that the user has sufficient understanding of the computer he or she is using to start up MATLAB. In later chapters we will assume that the user knows how to use an editor. The Macintosh version of MATLAB has a built-in editor, but the PC-Windows version and the X-windows version available on UNIX machines require the user to use a separate editor. It will be left to the readers to learn how to use an editor in their own environment. However, an editor will not be needed until we get to Chapter 3.

At any rate we will assume that you have started up MATLAB, and that you are now faced with a window on your computer which contains the MATLAB prompt, >>, and a cursor waiting for you to do something. This is called the MATLAB Command Window, and it is time to begin.

Numerical expressions

In its most elementary use, MATLAB is an extremely powerful calculator, with many built-in functions, and a very large and easily accessible memory. Let's start at the very beginning. Suppose you want to calculate a number such as $12.3(48.5 + \frac{342}{39})$. You can accomplish this using MATLAB by entering `12.3*(48.5+342/39)`. Try it. You should get the following:

```
>> 12.3*(48.5+342/39)

ans =

   704.4115
```

Notice that what you enter into MATLAB does not differ greatly from what you would write on a piece of paper. The only changes from the algebra that you use every day are the different symbols used for the algebraic operations. These are standard in the computer world, and are made necessary by the unavailability of the standard symbols on a keyboard. Here is a partial list of symbols used in MATLAB.

+	addition
−	subtraction
*	multiplication
/	right division
\	left division
^	exponentiation

While + and − have their standard meanings, * is used to indicate multiplication. You will notice that division can be indicated in two ways. The fraction $\frac{2}{3}$ can be indicated in MATLAB as either 2/3 or as 3\2. These are referred to as right division and left division respectively. Exponentiation is quite different in MATLAB; it has to be, since MATLAB has no way of entering superscripts. Consequently, the power 4^3 must be entered as 4^3. Consider

```
>> 0^0

ans =


        1
```

Are you surprised?

The order in which MATLAB performs arithmetic operations is exactly that taught in high school algebra courses. Exponentiations are done first, followed by multiplications and divisions, and finally by additions and subtractions. The standard order of precedence of arithmetic operations can be changed by inserting parentheses. For example, the result of 12.3*(48.5+342)/39 is quite different than the similar expression we computed earlier, as you will discover if you try it.

MATLAB allows the assignment of numerical values to variable names. For example, if you enter x = 3, then MATLAB will remember that x stands for 3 in subsequent computations. Therefore, computing 2.5*x will result in $2.5x = 2.5 \times 3 = 7.5$. You can also assign names to the results of computations. For example y = (x + 2)^x will result in y being given the value $(3 + 2)^3 = 125$.

You will have noticed that if you do not assign a name for a computation, MATLAB will assign the default name ans to the result. This name can always be used to refer to the results of the previous computation. For example:

```
>> 2+3

ans =

        5
```

```
>> ans/5

ans =

     1
```

MATLAB has a number of preassigned variables or constants. The constant $\pi = 3.14159...$ is given the name pi. The square root of -1 is given the names i and j, since each of these symbols is used by different people. There is no symbol for e, the base of the natural logarithms, but this can be easily computed as e = exp(1).

Mathematical functions

There is a long list of mathematical functions that are built into MATLAB. Included are all of the functions that are standard in calculus courses. Here is a list of MATLAB symbols, and their equivalents.

Elementary functions

abs(x)	The absolute value of x, i.e. $|x|$.
sqrt(x)	The square root of x, i.e. $\sqrt{x}$.
sign(x)	The signum of x, i.e. 0 if $x = 0$, -1 if $x < 0$, and $+1$ if $x > 0$.

The trigonometric functions

sin(x)	The sine of x.
cos(x)	The cosine of x.
tan(x)	The tangent of x.
cot(x)	The cotangent of x.
sec(x)	The secant of x.
csc(x)	The cosecant of x.

The inverse trigonometric functions

asin(x)	The inverse sine of x, i.e. $\arcsin(x)$.
acos(x)	The inverse cosine of x, i.e. $\arccos(x)$.
atan(x)	The inverse tangent of x, i.e. $\arctan(x)$.
acot(x)	The inverse cotangent of x, i.e. $\cot^{-1}(x)$.
asec(x)	The inverse secant of x, i.e. $\sec^{-1}(x)$.
acsc(x)	The inverse cosecant of x, i.e. $\csc^{-1}(x)$.

The exponential and logarithm functions

`exp(x)`   The exponential of x, i.e. e^x.
`log(x)`   The natural logarithm of x.
`log10(x)` The logarithm of x to base 10.

The hyperbolic funtions

`sinh(x)`   The hyperbolic sine of x.
`cosh(x)`   The hyperbolic cosine of x.
`tanh(x)`   The hyperbolic tangent of x.
`coth(x)`   The hyperbolic cotangent of x.
`sech(x)`   The hyperbolic secant of x.
`csch(x)`   The hyperbolic cosecant of x.

The inverse hyperbolic functions

`asinh(x)` The inverse hyperbolic sine of x, i.e. $\sinh^{-1}(x)$.
`acosh(x)` The inverse hyperbolic cosine of x, i.e. $\cosh^{-1}(x)$.
`atanh(x)` The inverse hyperbolic tangent of x, i.e. $\tanh^{-1}(x)$.
`acoth(x)` The inverse hyperbolic cotangent of x, i.e. $\coth^{-1}(x)$.
`asech(x)` The inverse hyperbolic secant of x, i.e. $\mathrm{sech}^{-1}(x)$.
`acsch(x)` The inverse hyperbolic cosecant of x, i.e. $\mathrm{csch}^{-1}(x)$.

For a more extensive list of the functions available, see the MATLAB *User's Guide,* or the MATLAB *Reference Guide.* All of these functions can be entered at the MATLAB prompt either alone or in combination. For example, to calculate $\sin(x) - \log\cos(x)$, where $x = 6$, we simply enter

```
>> x=6

x =

     6

>> sin(x)-log(cos(x))

ans =

    -0.2388
```

Output format

Up to now we have let MATLAB repeat everything that we enter at the prompt. Sometimes this is not useful. To prevent MATLAB from echoing what we tell it, simply enter a semicolon at the end of a command. For example, if we enter >> q=7; and then ask MATLAB what it thinks q is by entering >> q , we get the response

```
q =

    7
```

If you use MATLAB to compute $\cos(\pi)$, you get

```
>> cos(pi)

ans =

   -1
```

In this case MATLAB is smart enough to realize that the answer is an integer and it displays the answer in that form. However cos(3) is not an integer, and MATLAB gives us -0.9900 as its value. Thus if MATLAB is not sure that a number is an integer, it displays five significant figures in its answer. As another example, 1.57 is very close to $\pi/2$, and $\cos(\pi/2) = 0$. MATLAB gives us

```
>> cos(1.57)

ans =

   7.9633e-04
```

This is an example of MATLAB's exponential, or scientific notation. It stands for 7.9633×10^{-4}. In this case MATLAB again displays five significant digits in its answer. All of these illustrate the default format, which is called the short format. It is important to realize that although MATLAB only displays five significant digits in the default format, it is computing the answer to an accuracy of sixteen significant figures.

There are several other formats. We will discuss two of them. If it is necessary or desirable to have more significant digits displayed, enter format long at the MATLAB prompt. Then about sixteen significant digits will be displayed. Enter format long, cos(1.57) to see the difference.

There is another output format which we will find useful. If you enter format rat, then all numbers will be shown as rational numbers. This is called the rational format. If the numbers are actually irrational, MATLAB will find a very close rational approximation to the number. The rational format is most useful when you are working with numbers you know to be rational.

After using a different format, you can return to the standard, short format by entering `format short`.

Complex arithmetic

One of the nicest features of MATLAB is that it works with complex numbers as readily as it does with real numbers, and it does complex arithmetic with ease. To enter the complex number $z = 2 - 3i$ into MATLAB, we enter it the way it is written.

```
>> z=2-3i

z =

     2   -   3i
```

Then if we enter $w = 3 + 5i$, we can calculate sums, products and quotients of these numbers in exactly the same way we do for real numbers. For example,

```
>> w=3+5i;
>> z*w

ans =

   21.0000 + 1.0000i
```

and

```
>> z/w

ans =

   -0.2647 - 0.5588i
```

Any of the arithmetic functions listed earlier can be applied to complex numbers. For example,

```
>> y=sqrt(w)

y =

    2.1013 + 1.1897i
```

and

```
>> y*y

ans =

   3.0000 + 5.0000i
```

Since $y^2 = w$, it is a square root of the complex number w. The reader might try `cos(w)` and `exp(w)`. In particular, the reader might wish to verify Euler's formula

$$e^{i\theta} = \cos(\theta) + i \sin(\theta)$$

for several values of θ, including $\theta = 2\pi, \pi, \pi/2$.

The ease with which MATLAB handles complex numbers has one drawback. There is at least one case where the answer is not the one we expect. Use MATLAB to calculate $(-1)^{1/3}$. Most people would expect the answer -1, but MATLAB gives us

```
>> (-1)^(1/3)

ans =

   0.5000 + 0.8660i
```

At first glance this may seem strange, but if you cube this complex number you do get -1. Consequently MATLAB is finding a *complex* cube root of -1, while we would expect a real root. The situation is even worse, since in most of the cases where this will arise in this manual, it is not the complex cube root we want. We will want the cube root of -1 to be -1.

However, this is a price we have to pay for other benefits. For MATLAB to be so flexible that it can calculate roots of arbitrary order of arbitrary complex numbers, it is necessary that it should give what seems like a strange answer for the cube root of negative numbers. In fact the same applies to any odd root of negative numbers. What we need is a way to work around the problem.

Notice that if $x < 0$, then $x = -1 \times |x|$, and we can find a negative cube root as $-1 \times |x|^{1/3}$. Here we are taking the real cube root of the positive number $|x|$, and MATLAB does that the way we want it done. But suppose the situation arises where we do not know beforehand whether x is positive or negative. What we want is:

$$x^{1/3} = \begin{cases} |x|^{1/3}, & \text{if } x > 0; \\ 0, & \text{if } x = 0; \\ -1 \times |x|^{1/3}, & \text{if } x < 0. \end{cases}$$

7

To write this more succinctly we use the *signum* function sgn(x) (in MATLAB it is denoted by `sign(x)`). This function is defined to be

$$\text{sgn}(x) = \begin{cases} 1, & \text{if } x > 0; \\ 0, & \text{if } x = 0; \\ -1, & \text{if } x < 0. \end{cases}$$

Thus in all cases we have $x = \text{sgn}(x)\,|x|$, and the real cube root is

$$x^{1/3} = \text{sgn}(x)\,|x|^{1/3}.$$

In MATLAB, we would enter `sign(x)*abs(x)^(1/3)`.

Recording your work

It is frequently useful to be able to record what happens in a MATLAB session. For example, in the process of preparing a homework submission, it should not be necessary to copy all of the output from the computer screen. We ought to be able to do this automatically. The MATLAB `diary` command makes this possible.

For example, suppose you are doing your first homework assignment and you want to record what you are doing in MATLAB. To do this, choose a name, perhaps `hw1`, for the file in which you wish to record the output. Then enter `diary hw1` at the MATLAB prompt. From this point on, everything that appears in the Command Window will also be recorded in the file `hw1`. When you want to stop recording enter `diary off`. If you want to start recording again, enter `diary on`.

The file that is created is a simple text file. It can be opened by an editor or a word processing program and edited to remove extraneous material, or to add your comments. You can also print the file to get a hard copy.

A first order ordinary differential equation has the form

$$x' = f(t, x).$$

To solve this equation we must find a function $x(t)$ such that

$$x'(t) = f(t, x(t)) \qquad \text{for all } t.$$

This means that at every point $(t, x(t))$ on the graph of x, the graph must have slope equal to $f(t, x(t))$.

We can turn this interpretation around to give a geometric view of what a differential equation is, and what it means to solve the equation. At each point (t, x), the number $f(t, x)$ represents the slope of a solution curve through this point. Imagine, if you can, a small line segment attached to each point (t, x) with slope $f(t, x)$. This collection of lines is called a *direction line field*, and it provides the geometric interpretation of a differential equation. To find a solution we must find a curve in the plane which is tangent at each point to the direction line at that point.

Admittedly, it is difficult to visualize such a direction field. This is where the MATLAB routine `dfield` demonstrates its value.* Given a differential equation, it will plot the direction lines at a large number of points — enough so that the entire direction line field can be visualized by mentally interpolating between the field elements. This enables the user to get some geometric insight into the solutions of the equation. In addition the program will calculate and plot the solution through an initial point provided by the user. It has other options that will be explained later.

Starting `dfield`

To see `dfield` in action enter `dfield` at the MATLAB prompt. After a short wait, a new window will appear with the label **DFIELD Setup**. Figure 2.1 shows how this window looks on a UNIX machine. The appearance will differ slightly depending on your computer, but the functionality will be the same on all machines.

The **DFIELD Setup** window is an example of MATLAB's *Figure Window*. A Figure Window can assume a variety of forms as will soon become apparent. In a MATLAB session

---

* The MATLAB function `dfield` is not distributed with MATLAB. To discover if it is installed properly on your computer enter `help dfield` at the MATLAB prompt. If it is not installed, see the Preface for instructions on how to obtain it.

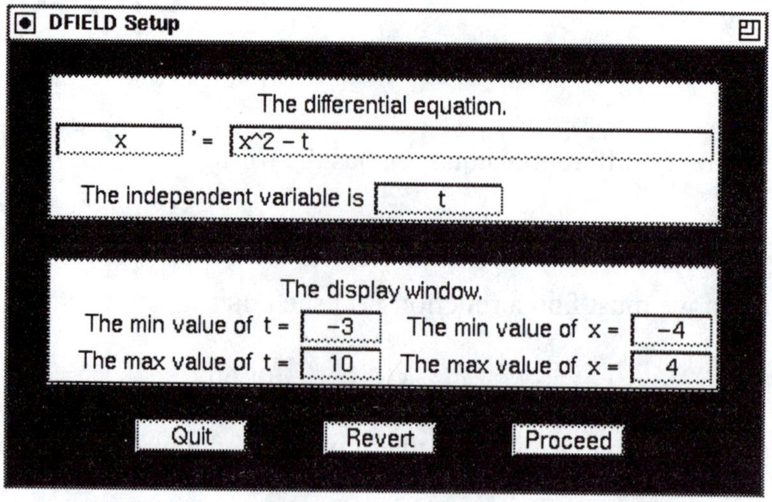

Figure 2.1. The Setup window for `dfield`.

there will always be one Command Window open on your screen, and perhaps a number of Figure Windows as well.

You will notice that the equation $x' = x^2 - t$ is already entered in the upper part of the **DFIELD Setup** window. In the middle a "display window" is described, in which the independent variable t is to satisfy $-3 \leq t \leq 10$, and the dependent variable x is to satisfy $-4 \leq x \leq 4$. At the bottom there are three buttons.

We will describe this window in detail later, but for now click the button with the label $\boxed{\textbf{Proceed}}$. After a few seconds another window will appear, this one labeled **DFIELD Display**. An example of this window is shown in Figure 2.2.

The most prominent feature of the **DFIELD Display** window is a rectangle labeled with the differential equation $x' = x^2 - t$ on the top, the independent variable t on the bottom, and the dependent variable x on the left. The dimensions of this rectangle are slightly larger than the rectangle specified in the **DFIELD Setup** window. Inside this rectangle the **DFIELD Display** window shows the direction field for the equation $x' = x^2 - t$. There is a grid of points, 20 in each direction, for a total of 400 points. At each such point with coordinates (t, x) there is shown a small line segment centered at (t, x) with slope equal to $x^2 - t$.

There are a pair of buttons on the **DFIELD Display** window, labeled $\boxed{\textbf{Quit}}$ and $\boxed{\textbf{Print}}$. There is a new menu labeled **DFIELD Options**. On a UNIX machine the menu appears in the upper left hand corner of the **DFIELD Display** window, as shown in Figure 2.2. On a Macintosh or in PC-Windows it appears on the menu bar.

In addition, below the direction field there is a message window through which `dfield` will communicate with the user. At this time it should contain the single word "Ready," indicating that it is ready to follow orders.

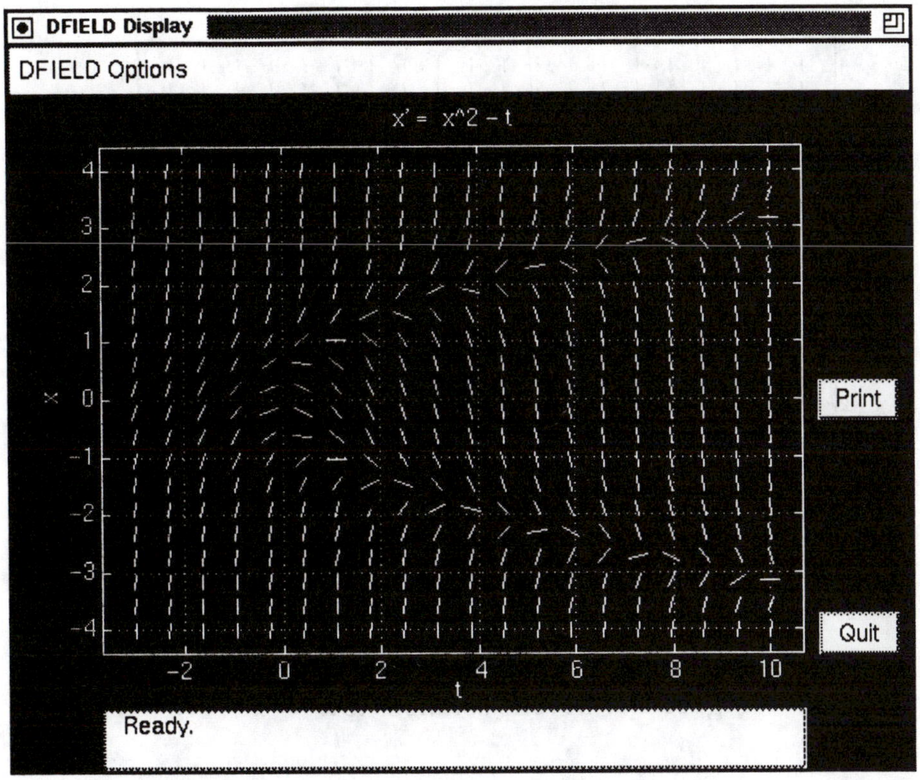

Figure 2.2. The display window for dfield.

A *solution curve* of a differential equation is the graph of a function which solves the equation. Computing and plotting a solution curve is very easy using dfield. Choose an initial point for the solution, move the mouse to that point, and click the mouse button. The computer will compute and plot the solution through the selected point, first in the direction in which the independent variable is increasing, and then in the opposite direction.

After computing and plotting a couple of solutions, the display should look something like Figure 2.3.

Changing the differential equation — using the DFIELD Setup window

The program dfield can be used to study any single first order differential equation. Notice that the **DFIELD Setup** window has three parts. The upper part allows the user to describe a differential equation. There are three text edit boxes, and the equation is entered into these boxes just the way it is written on paper, except that the MATLAB notation for the arithmetic operations must be used (see Chapter 1).

11

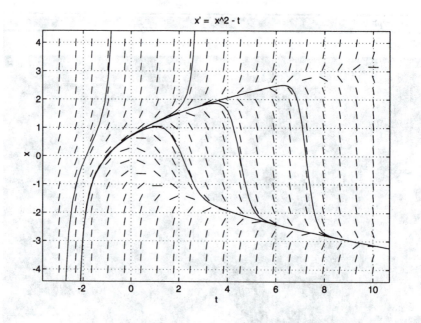

Figure 2.3. Several solutions to $x' = x^2 - t$.

To change what is in one of the boxes, move the mouse to that box and click once. A flashing line, called the text cursor, will appear in the box. The contents can now be changed in the way that is standard for your computer. Letters are entered from the keyboard, they are deleted by using the delete or backspace key, and the cursor can be moved backwards or forwards using the cursor motion keys. If you click twice or three times in the box, parts or all of the existing text will be highlighted, and will then be erased if something is entered to replace it. In some operating systems the tab key can be used to move from textbox to textbox.

To be more precise, to enter a differential equation

$$x' = \frac{dx}{dt} = f(t, x)$$

you enter the dependent variable, x in this case, into the first text edit box, the expression for $f(t, x)$ into the second, and the independent variable, which in this case is t, into the box with that label. The **DFIELD Setup** window opens with the default equation $x' = x^2 - t$ entered as an example.

As another example, to enter the equation $\frac{dy}{dx} = y - \cos(x)$, the upper part of the **DFIELD Setup** window must look like Figure 2.4.

In the differential equation $y' = f(t, y)$, the function $f(t, y)$ is called the *derivative function.* The expression for f must be entered into the second text edit box. It can be any function that can be expressed using algebra and the functions described in Chapter 1.

12

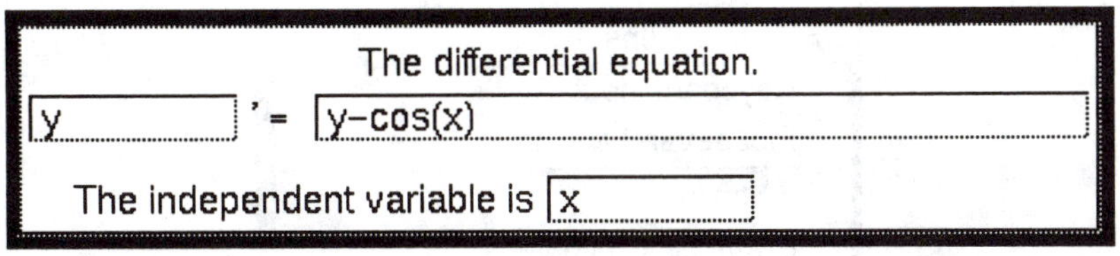

The differential equation.

y ' = y-cos(x)

The independent variable is x

Figure 2.4. The equation $y' = y - \cos(x)$ in the **DFIELD Setup** window.

The middle of the **DFIELD Setup** window is used to describe the rectangle on which the direction field should be displayed. If you compare the information entered here, with the dimensions of the rectangle displayed in the **DFIELD Display** window, you will see that this rectangle is slightly larger. It is just a little larger so that the direction lines can be displayed on the rectangle defined in the **DFIELD Setup** window. When entering a differential equation into the **DFIELD Setup** window, you should also enter the limits of the display rectangle you want into the middle portion of the **DFIELD Setup** window.

At the very bottom of the **DFIELD Setup** window there are three buttons. We have already seen that the ⎢ **Proceed** ⎢ button transfers the information in the **DFIELD Setup** window to the **DFIELD Display** window, and starts the computation of the direction field. Clicking the ⎢ **Revert** ⎢ button will return all of the settings in the **DFIELD Setup** window to what they were when you began to make changes. This is useful, for example, when you have made a number of changes and you decide that it would be easiest to start over. The ⎢ **Quit** ⎢ button should be used whenever you want to stop using dfield. It will close all related windows, and keep MATLAB running.

Replotting solutions

The user will have noticed that after dfield has plotted a solution step-by-step, the entire display window is replotted. This is necessary because, while the step-by-step solution is present on the screen, it is not a part of the picture internally to the program. The solution must be replotted before the display window is printed or copied, or the solution will be missing from the result.

Nevertheless, the replotting takes time. For that reason, immediate replotting can be disenabled, leaving the timing of replotting in the hands of the user. If the **Replot solutions later** option is chosen from the **DFIELD Options** menu (see Figure 2.5), subsequent solutions will not be replotted immediately. After the first such solution, a new button, labeled ⎢ **Replot** ⎢ will appear on the **DFIELD Display** window. Clicking this button will replot all solutions that need it. Notice that the ⎢ **Replot** ⎢ button is visible only if there is data that needs replotting. For example, after it is clicked, it disappears. There are other options that cause all of the solutions to be replotted, so the ⎢ **Replot** ⎢ button will sometimes disappear unexpectedly.

13

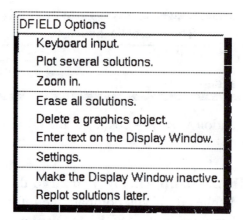

Figure 2.5. The options menu.

Immediate replotting can be reenabled by choosing the same option from the **DFIELD Options** menu. It is now labeled **Replot solutions immediately**.

Printing, quitting, and using clipboards

The easiest way to print the display window is to click the $\boxed{\textbf{Print}}$ button. This will replot the solutions that need it, and issue the print command.

If you have previous experience using MATLAB, you realize that the display window, or more generally, the current Figure Window, can be printed by entering `print` at the MATLAB prompt in the Command Window. In addition, in the Macintosh and the PC-Windows versions of MATLAB, the display window can be printed using methods that are standard to those operating systems. Before doing so you should ensure that all solutions have been replotted by clicking the $\boxed{\textbf{Replot}}$ button, if it is visible. The same caution applies if you want to use the MATLAB option of saving the display window to a postscript file. After clicking the $\boxed{\textbf{Replot}}$ button this can be accomplished from the Command Window in the usual way.

Sometimes MATLAB is very slow in arranging for a display window to be printed. While this is being done, strange things may happen to the window, such as the buttons disappearing. None of this is worrisome, unless you are really in a hurry. There is nothing you can do about it, so this is a time to be patient. Always wait until the word "Ready" appears in the message window before you try to do anything else with MATLAB. If you do not wait the results are unpredictable.

In the Macintosh and PC-Windows version of MATLAB, the contents of the display window can be copied into a clipboard, and from there into other documents in the standard ways. It is imperative that the $\boxed{\textbf{Replot}}$ button be clicked before copying to a clipboard, or solutions curves may be missing from the copy.

14

When you want to quit `dfield`, the best way is to use the [**Quit**] buttons found on the **DFIELD Setup** or on the **DFIELD Display** windows. Either of these will close all of the `dfield` windows in an orderly manner, and it will delete the temporary files that `dfield` creates in order to do its business.

If on occasion you do not quit `dfield` before quitting MATLAB, the temporary files can accumulate. These will have names like `dftp5678.m`. It is safe to delete these files, as long as you are not using `dfield` at the same time.

Other ways of choosing initial data

While choosing the initial point for a solution curve with the mouse is very convenient, sometimes it is necessary to choose the initial point with great accuracy. That is difficult to accomplish with the mouse. Instead, there is a **Keyboard input** option in the **DFIELD Options** menu (see Figure 2.5). If you select this option, another window will open labeled **DFIELD Keyboard input** (see Figure 2.6). Using this window, very accurate initial conditions can be entered. Clicking on the [**Compute**] button will start the computation with the initial conditions entered.

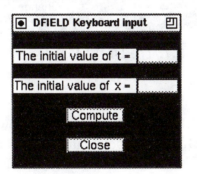

Figure 2.6. The Keyboard input window.

The second option in the **DFIELD Options** menu is useful when you want to choose several initial points at once. Choosing **Plot several solutions** allows you to select as many points with the mouse as you like, ending by hitting the **Return** key. All of the selected solution curves will be plotted.

Changing the size and appearance of the display window

Some people prefer to use a *vector field* rather than a direction field in the **DFIELD Display** window. In a vector field, a vector is attached to each point instead of the line segment used in a direction field. The vector has its base at the point in question, its direction is the slope, and the length of the vector reflects the magnitude of the derivative.

To change the direction field in the **DFIELD Display** window to a vector field, choose **Settings** from the **DFIELD Options** menu. Another window will open titled **DFIELD Settings** (see Figure 2.7). At the very top there are three radio buttons that allow you to choose between a line field, a vector field, or no field at all.

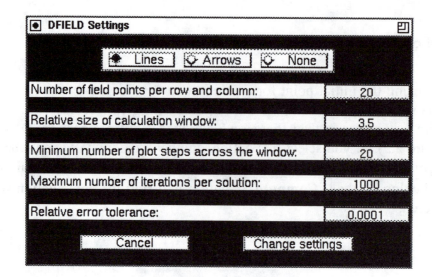

Figure 2.7. The DFIELD Settings window

The second line of the **DFIELD Settings** window allows the user to choose the number of field points displayed. The default is 20 points in each row and in each column. Any integer between 5 and 40 will be acceptable. The other settings which can be changed in the **DFIELD Settings** window will be explained later.

Choosing the **Zoom in** option from the **DFIELD Options** menu provides a convenient way to focus your attention on a portion of the current display window. After making this selection, you can choose a smaller display rectangle by clicking on two opposite corners of the new rectangle with the mouse.

There is no **Zoom out** option, since this can be accomplished by entering the description of a new display rectangle in the **DFIELD Setup** window. However, you must be careful not to change any of the information about the differential equation. If such a change is made, even if it is changed back to the original, `dfield` assumes that you want to start over, and it will erase all of the previously computed information.

Starting with version 4.2, MATLAB has its own `zoom` command. You can find out how to use it using MATLAB's help feature (enter `help zoom`). However, if you use this you will find that you can no longer plot solutions by choosing initial points in the Display Window.

16

This can be corrected by choosing **Make the Display Window inactive** from the **DFIELD Options** menu, and then selecting it again. The second time the item will read **Make the Display Window active**.

Personalizing the display window

Sometimes when you are preparing a display window for printing, you plot a solution curve you wish were not there. There are ways to correct for this. There are two choices on the **DFIELD Options** menu which allow you to erase items from the display. The **Erase all solutions** option is self explanatory.

The second one, **Delete a graphics object**, is much more flexible. It will allow you to delete any solution curve. It is only necessary to choose this option, and then to select the solution curve by clicking the mouse anywhere on the solution curve. It is best to choose a place on the solution curve which is as far away as possible from other curves or even line field elements. It is very easy to confuse `dfield`, in which case the wrong curve, or even a direction field element might be deleted instead of the curve you want.

As indicated in the previous paragraph, it is possible to delete a direction field element on purpose. It is possible using this option to delete any graphics object on the display window, including text.

There are three text elements which are part of the display window. These are the `title` at the top, the `xlabel` at the bottom, and the `ylabel` at the left. These are given default values by `dfield` using the information entered into the **DFIELD Setup** window, but they can be changed at any time. Suppose, for example, you have plotted several solutions of the equation $x' = x^2 - t$, and you decide that you really want the independent variable to be x, and the dependent variable to be y, so that the equation is $y' = y^2 - x$. Here are the commands to enter in the Command Window to change the title and the axis labels to reflect your desires:

```
>> xlabel('x')
>> ylabel('y')
>> title('y'' = y^2 - x')
```

Notice that each of the three commands takes as a parameter a text string which is contained between two single quotes. You can use any text string you think is appropriate.

There is a special problem with the title string in this example. You might think that it should read `'y' = y^2 - x'`. The problem is that the prime `'` is being used both to denote the derivative, and to indicate the beginning or end of the text string. When MATLAB reads the second prime in `'y'` it thinks this should be the end of the text string. Consequently, MATLAB needs a different way to designate a prime internal to a text string, and it uses the double prime (`''`) to do just this.

17

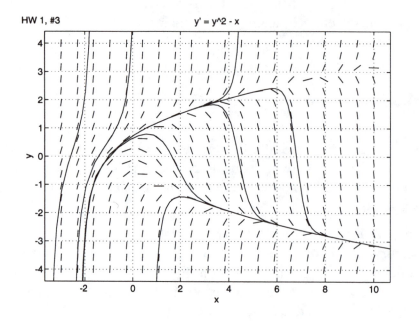

Figure 2.8. The altered display window.

It is also possible to add text at arbitrary points in the display window by choosing **Enter text on the Display Window** from the **DFIELD Options** menu. You will be prompted to enter the desired text in the Command Window. Suppose you want to label the display window with the homework assignment and number. The Command Window dialog would be:

```
>> Enter the text here. >  HW 1, #3
```

Now when the mouse cursor is in the display window it changes from the normal arrow to a cross. Position the cross at the point where you want the lower left part of the text to appear. Then click the mouse button, and the text will appear. You will probably want to choose a position for the text which is not in the direction field, since text on top of the direction field will be difficult to read. A typical result of this can be seen in Figure 2.8.

It can easily happen that your placement of the text does not please you. If so, you can remove the text using the **Delete a graphics object** option. When you are prompted to choose a graphics object, just click in the middle of the offending text.

Computational speed, accuracy, and kinky plots

The **DFIELD Settings** window gives the user options that affect the speed and accuracy of computation, and the appearance of the solution curves. All of these involve tradeoffs. The

DFIELD Settings window can be opened by choosing **Settings** from the **DFIELD Options** menu. We have already discussed the first two settings.

The design of dfield includes the definition of two windows. We have already discussed the **DFIELD Display** window. When you start dfield, the calculation window is 3.5 times as large as the display window in each dimension. The computation of a solution will stop only when the solution curve leaves the calculation window. This allows some room for zooming to a larger display window without having incomplete solution curves. It also allows for some reentrant solution curves, i.e. those which leave the display window at the top or the bottom and later return to it.

The third item in the **DFIELD Settings** window controls the relative size of the calculation window. It can be given any value bigger than or equal to 1. Clearly the smaller this number, the faster dfield will compute solutions, and the more likely that reentrant solutions will be lost. If you have a slower computer, and if you are not going to be zooming to a larger display window, it is not completely unreasonable to set this value to 1, but you have to realize that with this choice all reentrant solutions will be lost. A better choice would be 2 or 2.5. The default value of 3.5 seems to meet most needs.

dfield computes solutions using a numerical method that is closely related to those we will discuss in Chapter 5. It finds points on the solution curve, and it connects them with straight lines. If the points are too far apart, the curve will look kinky. The distance in the independent variable between two points is called the *step size*.

Within dfield the step size is controlled by two parameters which can be set in the **DFIELD Settings** window. The first is the **Minimum number of plot steps across the window.** If we call this number N, then the step size is at most $1/N$ times the width of the display window. The second is called the **Relative error tolerance**. If we denote this number by T, then, roughly speaking, the step size is chosen to ensure that the error being made at any step is less than T times the absolute value of the solution.

The larger N is, and the smaller T is, the more accurate the solution. On the other hand, the number of steps is increased, and as a result the computation is slower. N can be any non-negative integer. When N is small, even if the tolerance T is small, the step size can be large and the solution curves may have kinks. As you can see the trade off here between computation speed and the appearance of the solution curves is quite complicated.

The default settings should be sufficient in most cases. If you are using a very fast computer, you should consider setting the tolerance T to be equal to 10^{-5} or 10^{-6}, i.e. 1e-5 or 1e-6 in MATLAB's parlance. If your computer is not as fast, and the speed with which solutions are being computed bothers you, set $N = 0$, and perhaps even $T = 0.001$. You can then decrease T as you feel necessary to increase your accuracy, and to make the solution curves smoother.

The final setting in the **DFIELD Settings** window is the **Maximum number of iterations per solution**. This is the maximum number of steps that dfield will compute in finding a solution. The default value is 1000. There are very few solution curves that will require this many steps. If dfield uses this many steps, it will stop computing and so inform you by a

19

message in the Command Window. If this happens, and it is clear from the display that `dfield` has not finished computing, you can increase this number and try again.

A rarely needed option

It may happen that you want to click the mouse while its cursor is in the display window, without a solution curve being plotted. If this is the case, choose **Make the Display Window inactive** from the **DFIELD Options** menu. When you want to plot more solution curves choose this option again — it is now labeled **Make the Display Window active.** It is not necessary to make the display window inactive before using any of the `dfield` options.

Using MATLAB while `dfield` is open

This is certainly possible. However, there is an avoidable problem that arises if you try to plot something while `dfield` is open. You may discover that you are plotting to one of the `dfield` windows, right on top of what is already there. Sometimes this is a useful feature, but other times it is not. To avoid this, enter `figure` at the MATLAB prompt before you plot for the first time. A new Figure Window will open with nothing in it, and with a title like **Figure No. nn**, where nn is a number on the order of 3 or 4.

Future plotting commands will be directed to this window, as long as you do not click the mouse while the cursor is in another window. If, by chance, you should select another window in this way, you can change back to the right Figure Window, either by clicking the mouse in the correct Figure Window, or by entering `figure(nn)` at the MATLAB prompt.

All is not lost if you mistakenly plot something to the **DFIELD Display** window. Simply use the **Delete a graphics object** option from the **DFIELD Options** menu to remove the offending plots.

The logistic equation

As an example of the use of `dfield`, let's examine the logistic equation $p' = \alpha p(1 - p/N)$. Here p is a population and t is time. Frequently, this equation is normalized by introducing the *normalized population* $q = p/N$. When written in terms of q the logistic equation becomes $q' = \alpha q(1 - q)$. We call this the *normalized logistic equation.*

Let's choose $\alpha = 0.1$, and enter the resulting equation, $q' = 0.1 q(1 - q)$, into the **DFIELD Setup** window. The independent variable is time, denoted by t. We will want time to be positive, and a little experimentation shows us that the interesting things happen before $t = 100$. Similarly, we expect any population to be positive, and experiment shows that the normalized population limits out at $q = 1$. (Actually, that's why it called the normalized

20

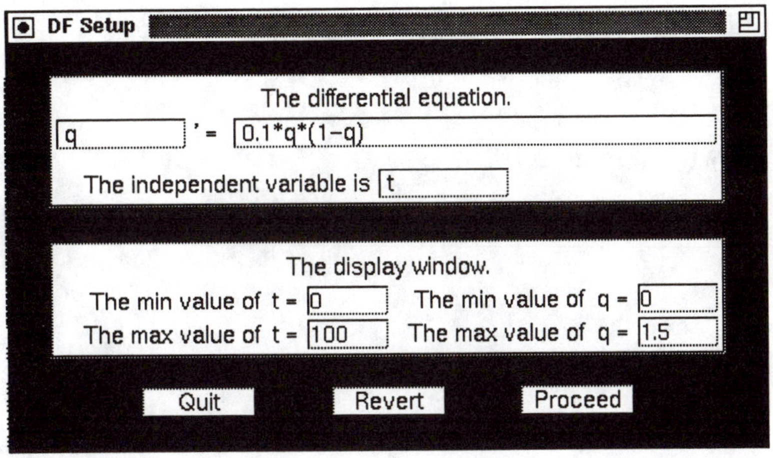

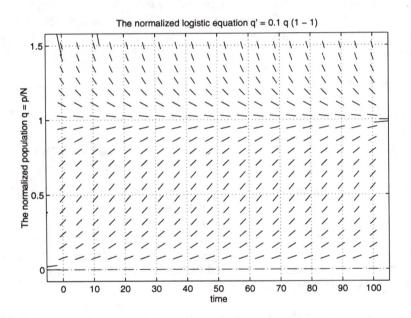

Figure 2.9. The setup window for the normalized logistic equation.

population.) Hence we define our display window by $0 \le t \le 100$, and $0 \le q \le 1.5$. The Setup Window with the correct information entered is shown in Figure 2.9.

Figure 2.10. Solutions of the normalized logistic equation.

After clicking ⬜ **Proceed** , and plotting a few solutions, we decide that the labels could be more descriptive. We, therefore, execute the following commands:

21

```
>> xlabel('time')
>> ylabel('The normalized population q = p/N')
>> title('The logistic equation q'' = 0.1 q (1 - q)')
```

The result is shown in Figure 2.10.

Exercises

1. This problem requires you to use **dfield** with the differential equation $y' = y^2 - t^2$.

   a) Print out the direction field for this differential equation with the display window defined by $t \in [-5, 5]$ and $x \in [-5, 5]$. On this printout sketch with a pencil as best you can the solution curves through the initial points $(t_0, y_0) = (0, 0), (-2, 0), (-3, 0), (0, 1)$, and $(4, 0)$. Remember that the solution curves must be tangent to the direction lines at each point.

   b) Use **dfield** to plot the same solution curves to check your accuracy. Turn in both versions.

2. This problem requires you to use **dfield** with the differential equation $y' = 2ty/(1 + y^2)$.

   a) Print out the direction field for this differential equation with the display window defined by $t \in [-4, 4]$ and $x \in [-4, 4]$. On this printout sketch with a pencil as best you can the solution curves through the initial points $(t_0, y_0) = (0, 0), (0, 2), (0, 1), (-2, 1)$, and $(-3, 1)$. Remember that the solution curves must be tangent to the direction lines at each point.

   b) Use **dfield** to plot the same solution curves to check your accuracy. Turn in both versions.

3. Use **dfield** to plot a few solution curves to the equation $x' + x \sin(t) = \cos(t)$. Use the display window defined by $x \in (-10, 10)$ and $t \in (-10, 10)$.

4. Use **dfield** to plot the solution curves for the equation $x' = 1 - t^2 + \sin(tx)$ with initial values $x = -3, -2, -1, 0, 1, 2, 3$ at $t = 0$. Find a good display window by experimentation.

5. Consider the differential equation
$$(1 + t^2)y' + 4ty = t.$$

   a) Use **dfield** to calculate and plot a few solutions with different initial points. (Use the display window defined by $t \in [-5, 5]$ and $y \in [-5, 5]$.) In particular plot the solution curve with initial point $y(1) = 1/4$ (use the **Keyboard input** option). Print out the Figure Window and turn it in as part of this assignment.

   b) What do you conjecture is the limiting behavior of the solutions of this differential equation as $t \to \infty$?

   c) Find the general analytic solution to this equation.

   d) Verify the conjecture you made in part b), or if you no longer believe it, make a new conjecture and verify that.

6. A certain lake has a volume of V km^3. It is fed by a river at a rate of r_i km^3/year, and there is another river which is fed by the lake at a rate which keeps the volume of the lake constant. In addition there

is a factory on the lake which introduces a pollutant into the lake at the rate of p km^3/year. This means that the rate of flow from the lake into the outlet river is $(p + r_i)$ km^3/year. Let $x(t)$ denote the volume of the pollutant in the lake at time t, and let $c(t) = x(t)/V$ denote the concentration of the pollutant.

a) Show that, under the assumption of immediate and perfect mixing of the pollutant into the lake water, the concentration satisfies the differential equation $c' + ((p + r_i)/V)c = p/V$.

b) In the previous problem suppose that $V = 100$, $r_i = 50$, and $p = 2$. Suppose that the factory starts operating at time $t = 0$, so that the initial concentration is 0. Use `dfield` to plot the solution.

c) It has been determined that a concentration of over 2% is hazardous for the fish in the lake. Approximately how long will it take until this concentration is reached? You can "zoom in" on the `dfield` plot to enable a more accurate estimate.

d) What is the limiting concentration? About how long does it take for the concentration to reach a concentration of 3.5%?

7. Suppose the factory in the previous problem stops operating at time $t = 0$, and that the concentration was 3.5% at that time. Approximately how long will it take before the concentration falls below 2%, and the lake is no longer hazardous for fish? Notice that $p = 0$ for this exercise.

8. Rivers do not flow at the same rate the year around. They tend to be full in the Spring when the snow melts, and to flow more slowly in the Fall. To take this into account, suppose the flow of our river is

$$r_i = 50 + 20\cos(2\pi(t - 1/3)).$$

Our river flows at its maximum rate one-third into the year, i.e., around the first of May, and at its minimum around the first of September.

a) Use `dfield` to plot the concentration for the factory in the previous two problems, and this flow rate. (You will undoubtedly have to play with the **DFIELD Settings** in order to get a smooth graph.) How would you describe the behavior of the concentration for large values of time? Compare the results for several choices of initial concentration between 0% and 4%.

b) It might be expected that after settling into a steady state, the concentration would be greatest when the flow was smallest, i.e., around the first of September. At what time of year does it actually occur?

9. Use `dfield` to plot several solutions to the equation $z' = (z - t)^{5/3}$. (**Hint:** Notice that when $z < t$, $z' < 0$, so the direction field should point down, and the solution curves should be decreasing. You might have difficulty getting the direction field and the solutions to look like that. If so read the section in Chapter 1 on complex arithmetic, especially the last couple of paragraphs.)

One of the properties of solutions of differential equations that is of interest is the asymptotic behavior of solutions for large values of the independent variable. When it is possible to compute the exact solution, this behavior can usually be determined. However, it is sometimes possible to determine this asymptotic behavior even if the exact solution cannot be found.

One case in which this is possible is for equations of the form $x' = f(x)$. These are equations for which the right-hand side does not depend on the independent variable, which we will denote by t. Such equations are called *autonomous* equations. If x_0 satisfies $f(x_0) = 0$, then $x(t) \equiv x_0$ is a solution to the equation. Such a point x_0 is called an *equilibrium point*. It can be shown that if $f'(x_0) < 0$, then every solution curve that has an initial value near x_0 converges to x_0 as $t \to \infty$. In this case x_0 is called a *stable* equilibrium point. If $f'(x_0) > 0$, then every solution curve that has an initial value near x_0 diverges away from x_0 as $t \to \infty$, and x_0 is called an *unstable* equilibrium point. If $f'(x_0) = 0$, no conclusion can be drawn about the behavior of solution curves. In this case the equilibrium point may fail to be either stable or unstable.

10. For each of the following autonomous equations find the equilibrium points in the indicated interval. Determine the type of each equilibrium point. Verify your results using dfield.

a) $x' = \cos(\pi x)$, $x \in [-3, 3]$.

b) $x' = x(x - 2)$, $-\infty < x < \infty$.

c) $x' = x(x - 2)^2$, $-\infty < x < \infty$.

d) $x' = x(x - 2)^3$, $-\infty < x < \infty$.

e) $x' = x(1 + e^{-x} - x^2)$, $-1 \leq x \leq 2$. In this case you will not be able to solve explicitly for all of the equilibrium points. Instead, turn the problem around. Use dfield to plot some solutions, and from that information calculate approximately where the equilibrium points are, and determine the type of each.

It is also possible to describe the asymptotic behavior of solutions for linear equations. First we will examine *homogeneous* linear equations. These are equations of the form $x' + a(t)x = 0$. If a is a constant, then the general solution is $y(t) = Ce^{-at}$. Hence, if a is a positive constant, every solution tends to 0 as $t \to \infty$. It is not hard to prove that if $a(t) \geq c > 0$ for all t, then the same thing is true; i.e. every solution to the homogeneous equation tends to 0 as $t \to \infty$.

Next consider the *inhomogeneous* equation $x' + a(t)x = A\cos(\omega t)$. Again, if a is a constant, we can solve the equation explicitly. The general solution is

$$y(t) = Ce^{-at} + \frac{A\omega}{a(a^2 + \omega^2)}(a\cos(\omega t) + \omega\sin(\omega t))$$

$$= Ce^{-at} + \frac{A\omega}{a\sqrt{a^2 + \omega^2}}\cos(\omega t - \delta),$$

where $\delta = \arctan(\omega/a)$. If a is a positive constant, then e^{-at} tends to 0, so every solution gets closer and closer to the particular solution

$$y_p(t) = \frac{A\omega}{a\sqrt{a^2 + \omega^2}}\cos(\omega t - \delta)$$

as $t \to \infty$. We will say, therefore, that all solutions are asymptotic to each other and to this particular solution as $t \to \infty$.

Once again, this result carries over to the variable coefficient case. If we assume that $a(t) \geq c > 0$ for all t, then every solution to the inhomogeneous equation described above is asymptotic to each other as $t \to \infty$. In this generality it is not possible to say too much more about the asymptotic solution. In particular, it is not possible to conclude that the asymptotic solution is periodic.

11. Using dfield, verify in each of the following cases that the solutions are asymptotic to each other as $t \to \infty$.

    a) $x' + (2 + \sin(2t))x = 0.$

    b) $x' + (2 + \sin(2t))x = 3\cos(3t).$

    c) $x' + (1 + e^{-t})x = 0.$

    d) $x' + (1 + e^{-t})x = 4\cos(\pi t).$

12. The logistic equation is

$$p' = \alpha\, p\left(1 - \frac{p}{K}\right).$$

The quantities α and K are the parameters of the equation. Usually the parameters are constants, and in that case the kind of analysis carried out in problem # 10 shows that for any solution $p(t)$ which has a positive initial value we have $p(t) \to K$ as $t \to \infty$.

Sometimes the parameters in the logistic equation are not really constants, but depend on time. Consider the equation in the following four cases, with $\alpha = 1$ in each case:

    i) $K(t) = 1,$

    ii) $K(t) = 1 - \frac{1}{2}e^{-t},$

    iii) $K(t) = 1 + t,$

    iv) $K(t) = 1 - \frac{1}{2}\cos(2\pi t).$

For each case you are to use dfield to plot solution curves with several choices of initial population p_0 between 0 and 3, and with $t_0 = 0$. In particular, you want to analyze the behavior of the solutions for large times (i.e, as $t \to \infty$). The first case is the standard logistic equation with parameters $\alpha = K = 1$. Consequently, whatever the initial population, we expect that $p(t) \to 1$ as $t \to \infty$. This case is here for comparison with the other three. What will happen in these other cases is not so clear, and it is your job to find out.

In cases ii) and iii), $K(t)$ is monotone increasing. In case ii), $K(t)$ is asymptotic to 1, and in case iii), $K(t)$ is unbounded, but of a very simple nature. These might model a situation of a human population where, due to technological improvement, the availability of resources is increasing with time, and therefore the effects of competition become less severe.

Case iv) is perhaps the most interesting. Here the parameter $K(t)$ is periodic in time with period 1, which might be considered to be one year. This might model a population of insects or small animals that are affected by the seasons. You will notice that the asymptotic behavior as $t \to \infty$ reflects the behavior of K. The solution does not tend to a constant, but nevertheless all solutions have the same behavior for large values of t. In particular you should take notice of the location of the maxima of K and of the steady state of p. You can use the "zoom in" option to get a better picture of this.

a) Use `dfield` to plot several solutions to each of the four equations. (It is up to you to find a display window that is appropriate to the problem at hand.)

b) For each case, based on the plot done in the previous problem, describe the asymptotic behavior of the solutions to the equation. In particular, compare this asymptotic behavior to the asymptotic behavior of K. It might be helpful to plot K on the same scale as the plot produced in part 1. In the first two cases the solutions will be asymptotic to a constant. In the other two the solutions will be asymptotic to a function. You are not expected to find that function explicitly, but you should be able to describe it qualitatively.

c) It is possible to find the solutions to these equations explicitly (except, perhaps, for the evaluation of an integral). Find these solutions. (**Hint:** Look up Bernoulli's equation in your textbook.)

A version of the existence and uniqueness theorem which is well adapted to the use of `dfield` is the following:

Theorem. *Suppose that the function $f(t, x)$ is defined in the rectangle R defined by $a \le t \le b$ and $c \le x \le d$. Suppose also that f and $\frac{\partial f}{\partial x}$ are both continuous in R. Then, given any point $(t_0, x_0) \in R$, there is one and only one function $x(t)$ defined for t in an interval containing t_0 such that $x(t_0) = x_0$, and $x' = f(t, x)$. Furthermore, the function $x(t)$ is defined both for $t > t_0$ and for $t < t_0$, at least until the graph of x leaves the rectangle R through one of its four edges.*

The rectangle R in the theorem is like the display window used by `dfield`. The theorem then says that through any point in R there is one and only one solution curve. This means that solution curves cannot cross each other, since otherwise there would be two solution curves through the point of intersection. Furthermore, this solution curve exists on each side of t_0 at least until it leaves R through any one of the four sides. Notice that it can leave through the top or bottom as easily as through the right and left sides.

The point of the next exercise is to illustrate limitations on the t-interval in which the solution exists which are not obvious in the statement of the theorem. Even for some very nice equations the solutions "blow up" in a finite amount of time (i.e., t). The important thing is that there is a solution defined in some interval containing t_0, although the precise interval in which the solution is defined might be quite different (perhaps smaller) than the t-interval in the definition of the rectangle R.

13. Consider the differential equation $y' = y^2$ with initial condition $y(t_0) = 1$.

   a) Find the exact solution. What is the largest t-interval containing t_0 in which the solution exists? You will discover that the solution blows up at some point.

   b) Use `dfield` to plot the solution curves for $t_0 = 0$, $t_0 = 1$, and $t_0 = 2$. Use the display window defined by $-1 \le t \le 3$ and $0 \le y \le 4$.

   c) In part a), we observe that each of these solutions blows up at a point in the interval $[-1, 3]$. Explain why this does not contradict the existence part of the theorem.

26

Despite the seeming generality of the uniqueness theorem, there are initial value problems which have more than one solution. This is illustrated by the next problem.

14. Consider the differential equation $y' = \sqrt{|y|}$. Notice that $y(t) \equiv 0$ is a solution with the initial condition $y(0) = 0$. (Of course by $\sqrt{|y|}$ we mean the **nonnegative** square root.)

   a) This equation is separable. Use this to find a solution to the equation with the initial value $y(t_0) = 0$ assuming that $y \geq 0$. The answer is $y(t) = (t - t_0)^2/4$. Notice, however, that this is a solution only for $t > t_0$.

   b) Show that the function
   $$y(t) = \begin{cases} 0, & \text{if } t \leq t_0; \\ (t - t_0)^2/4, & \text{if } t \geq t_0 \end{cases}$$
   is continuous, has a continuous first derivative, and satisfies the differential equation $y' = \sqrt{|y|}$.

   c) For any $t_0 \geq 0$ the function defined in part b) satisfies the initial condition $y(0) = 0$. Why doesn't this violate the uniqueness part of the theorem?

   d) Find another solution to the initial value problem in a) by assuming that $y \leq 0$.

   e) You might be curious (as was the author) about what `dfield` will do with this equation. Find out. Use the rectangle defined by $-1 \leq t \leq 1$ and $-1 \leq y \leq 1$ and plot the solution with initial value $y(0) = 0$. Also plot the solution for $y(0) = 10^{-50}$ (the MATLAB notation for 10^{-50} is `1e-50`). Plot a few other solutions as well. Do you see evidence of the non-uniqueness observed in part c)?

An important aspect of differential equations is the dependence of solutions on initial conditions. There are two points to be made.

First, we have a theorem which says that the solutions are contiuous with respect to the initial conditions. More precisely,

Theorem. *Suppose that the function $f(t, x)$ is defined in the rectangle R defined by $a \leq t \leq b$ and $c \leq x \leq d$. Suppose also that f and $\frac{\partial f}{\partial x}$ are both continuous in R, and that*

$$|\frac{\partial f}{\partial x}| \leq L \quad \text{for all } (t, x) \in R.$$

If (t_0, x_0) and (t_0, y_0) are both in R, and if

$$\begin{array}{ccc} x' = f(t, x) & & y' = f(t, y) \\ & and & \\ x(t_0) = x_0 & & y(t_0) = y_0 \end{array}$$

then for $t > t_0$

$$|x(t) - y(t)| \leq e^{L(t-t_0)}|x_0 - y_0|$$

as long as both solution curves remain in R.

Roughly, the theorem says that if we have initial values that are sufficiently close to each other, the solutions will remain close, at least if we restrict our view to the rectangle R. Since it is easy to make measurement mistakes, and thereby get initial values off by a little, this is reassuring.

For the second point, we notice that although the dependence on the initial condition is continuous, the term $e^{L(t-t_0)}$ allows the solutions to get exponentially far apart as the interval between t and t_0 increases. I.e., the solutions can still be extremely sensitive to the initial conditions, especially over long t inervals. The following exercise will illustrate this point.

15. Consider the differential equation $x' = x(1 - x^2)$.

   a) Verify that $x(t) \equiv 0$ is the solution with initial value $x(0) = 0$.

   b) Use dfield to find approximately how close the initial value y_0 must be to 0 so that the solution $y(t)$ of our equation with that initial value satisfies $y(t) \leq 0.1$ for $0 \leq t \leq t_f$, with $t_f = 2$. You can use the display window $0 \leq t \leq 2$, and $0 \leq x \leq 0.1$, and experiment with initial values in the **Keyboard input** window, until you get close enough. Do not try to be too precise. Two significant figures is sufficient.

   c) As the length of the t interval is increased, how close must y_0 be to 0 in order to insure the same accuracy? To find out, repeat part b) with $tf = 4, 6, 8$, and 10.

It is clear from the results of the last problem that the solutions can be extremely sensitive to changes in the initial conditions. This sensitivity allows chaos to occur in deterministic systems, which is the subject of much current research.

3. Vectors, Matrices, and Array Operations

A powerful feature of MATLAB is that every numerical quantity is considered to be a complex matrix! For those of you who do not already know, a *matrix* is a rectangular array of numbers. For example:

$$A = \begin{pmatrix} 1 & \pi & 0 \\ 0 & 4 & 5 \end{pmatrix}$$

is a matrix with 2 rows and 3 columns.

Matrices in MATLAB

If you want to enter the matrix **A** into MATLAB proceed as follows:

```
>> A=[1 pi 0;0 4 5];
>> A

A =

   1.0000    3.1416         0
        0    4.0000    5.0000
```

From this point on MATLAB will consider A to be this matrix.

The *size* of a matrix is the number of rows and columns. For example:

```
>> size(A)

ans =

   2    3
```

verifies that A has 2 rows and 3 columns. Two matrices are said to have the same size if they have the same number of rows and the same number of columns.

If you don't believe that MATLAB thinks that the number 5 is a matrix, try the following:

```
>> size(5)

ans =

   1    1
```

Thus MATLAB thinks that 5, or any other complex number, is a matrix with one row and one column. Enter `a=sqrt(-1),size(a)`.

A *vector* is a list of numbers.* It can be a vertical list, in which case it is called a *column vector*, or it can be a horizontal list, in which case it is called a *row vector*. Thus, vectors are special cases of matrices.

For example, we can define a column vector as follows:

```
>> v=[1;4;9;5]

v =

     1
     4
     9
     5
```

To enter a row vector is just as easy.

```
>> u=[sqrt(2),2,6,9]

u =

    1.4142    2.0000    6.0000    9.0000
```

Notice that we used commas instead of semicolons in this entry. In fact, between elements of the same row of a matrix we could just as well use spaces. For example, we could have entered the vector u by entering `u=[sqrt(2) 2 6 9]`. Similarly, the separation between rows can be

_____

* The word *vector* is one of the most overused terms in mathematics and its applications. To a physicist or a geometer, a vector is a directed line segment. To an algebraist or to many engineers, a vector is a list of numbers. To users of more advanced parts of linear algebra, a vector is an element of a vector space. In this latter, most general case, a vector could be any of the above examples, a polynomial, a more general function, or an example of, quite literally, any class of mathematical objects which can be added together and scaled by multiplication.

All too often the meaning in any particular situation is not explained. The result is very confusing to the student. When the word vector appears, a student should make a concerted effort to discover the meaning that is used in the current setting.

When using MATLAB, the situation is clear. A vector is a list of numbers, which may be complex.

indicated by carriage returns. The column vector v could have been entered as

```
>> v=[1
4
9
5]
```

Thus commas or spaces separate entries in a row, while semicolons or carriage returns separate rows.

The *length* of a vector is the number of elements in the list. For example, the length of each of u and v is 4. The MATLAB command length will disclose the length of any vector. Try length(u) and length(v).

Operations on matrices

If *A* and *B* are matrices of the same size, they can be added together. For example:

```
>> B=[4,0,0
12.5,-4,0]

B =

    4.0000         0         0
   12.5000   -4.0000         0

>> C=A+B

C =

    5.0000    3.1416         0
   12.5000         0    5.0000
```

You will notice that in the sum of two matrices, each element is the sum of the corresponding elements in the summands. The same is true for the difference of two matrices. Try C-A, and see what you get.

The operation of *matrix multiplication* is more complicated. We will start with the special case of multiplying a row vector with a column vector. As you will see, even the most complicated case comes down to this case in the end. It is required that the two have the same length, as do u and v. The product of two such vectors is defined to be the sum of the products of the corresponding elements. For example,

$$\mathbf{u} \cdot \mathbf{v} = (\sqrt{2} \quad 2 \quad 6 \quad 9) \cdot \begin{pmatrix} 1 \\ 4 \\ 9 \\ 5 \end{pmatrix}$$

$$= \sqrt{2} \cdot 1 + 2 \cdot 4 + 6 \cdot 9 + 9 \cdot 5$$

$$= 107 + \sqrt{2}.$$

In MATLAB we get

```
>> u*v

ans =

   108.4142
```

which is the same thing.

Next let's consider the product of a matrix and a column vector. Let

$$A = \begin{pmatrix} 3 & -2 & -4 & 3 \\ 0 & 5 & 2 & -1 \\ 0 & 4 & 9 & -4 \end{pmatrix}.$$

Such a matrix can be considered as a vertical list of row vectors. For example, we have

$$A = \begin{pmatrix} \mathbf{a}_1 \\ \mathbf{a}_2 \\ \mathbf{a}_3 \end{pmatrix},$$

where

$$\begin{aligned} \mathbf{a}_1 &= (3 \quad -2 \quad -4 \quad 3) \\ \mathbf{a}_2 &= (0 \quad 5 \quad 2 \quad -1) \\ \mathbf{a}_3 &= (0 \quad 4 \quad 9 \quad -4) \end{aligned}$$

If we interpret the matrix A as a vertical list of row vectors, the product of A with a column vector $\mathbf{v}$ is simply the vertical list of the products of the row vectors in A and the vector $\mathbf{v}$. In our specific example we get

$$\mathbf{A} \cdot \mathbf{v} = \begin{pmatrix} \mathbf{a}_1 \\ \mathbf{a}_2 \\ \mathbf{a}_3 \end{pmatrix} \cdot \mathbf{v}$$

$$= \begin{pmatrix} \mathbf{a}_1 \cdot \mathbf{v} \\ \mathbf{a}_2 \cdot \mathbf{v} \\ \mathbf{a}_3 \cdot \mathbf{v} \end{pmatrix}$$

$$= \begin{pmatrix} -26 \\ 33 \\ 77 \end{pmatrix}.$$

MATLAB does the multiplication very easily.

```
>> A=[3,-2,-4,3;0,5,2,-1;0,4,9,-4];
>> A*v

ans =

   -26
    33
    77
```

The general case of the product of two matrices **A** and **B** can be reduced to the previous case. Now we consider **B** to be a horizontal list of column vectors. For example, if

$$\mathbf{B} = \begin{pmatrix} 9 & -5 \\ 4 & -7 \\ 7 & 8 \\ 1 & -4 \end{pmatrix},$$

we have $\mathbf{B} = (\mathbf{b}_1 \ \mathbf{b}_2)$, where

$$\mathbf{b}_1 = \begin{pmatrix} 9 \\ 4 \\ 7 \\ 1 \end{pmatrix} \quad \text{and} \quad \mathbf{b}_2 = \begin{pmatrix} -5 \\ -7 \\ 8 \\ -4 \end{pmatrix}.$$

Then the product of **A** and **B** is the horizontal list of the products of **A** with the column vectors of **B**. In our specific example this comes down to

$$\mathbf{A} \cdot \mathbf{B} = \mathbf{A} \cdot (\mathbf{b}_1 \ \mathbf{b}_2) = (\mathbf{A} \cdot \mathbf{b}_1 \ \mathbf{A} \cdot \mathbf{b}_2) = \begin{pmatrix} -6 & -45 \\ 33 & -15 \\ 75 & 60 \end{pmatrix}.$$

As usual, MATLAB is very efficient.

```
>> B=[9 -5; 4 -7; 7 8; 1 -4]

B =

     9    -5
     4    -7
     7     8
     1    -4
```

```
>> A*B

ans =

    -6    -45
    33    -15
    75     60
```

If we unravel the definition of $\mathbf{A} \cdot \mathbf{B}$ one step, we see that, in our example,

$$\mathbf{A} \cdot \mathbf{B} = \begin{pmatrix} \mathbf{a}_1 \cdot \mathbf{b}_1 & \mathbf{a}_1 \cdot \mathbf{b}_2 \\ \mathbf{a}_2 \cdot \mathbf{b}_1 & \mathbf{a}_2 \cdot \mathbf{b}_2 \\ \mathbf{a}_3 \cdot \mathbf{b}_1 & \mathbf{a}_3 \cdot \mathbf{b}_2 \end{pmatrix}.$$

Thus the element in the ith row and the jth column of the product $\mathbf{A} \cdot \mathbf{B}$ is the product of the ith row vector of $\mathbf{A}$ and the jth column vector of $\mathbf{B}$.

This interpretation of matrix multiplication is valid, in general, and it gives the method by which matrix products are usually computed. Let's state this more precisely. Suppose that $\mathbf{A}$ is a matrix with p rows and q columns, and that $\mathbf{B}$ is a matrix with r rows and s columns. Then $\mathbf{A}$ can be considered as a vertical list of p row vectors, each of which has length q,

$$\mathbf{A} = \begin{pmatrix} \mathbf{a}_1 \\ \mathbf{a}_2 \\ \vdots \\ \mathbf{a}_p \end{pmatrix}.$$

Similarly, $\mathbf{B}$ can be considered as a horizontal list of s column vectors, each of length r

$$\mathbf{B} = (\mathbf{b}_1 \, \mathbf{b}_2 \, \ldots \mathbf{b}_s)$$

Then the element in the ith row and the jth column of the product $\mathbf{A} \cdot \mathbf{B}$ is the product of the ith row vector of $\mathbf{A}$ and the jth column vector of $\mathbf{B}$, i.e.

$$\mathbf{A} \cdot \mathbf{B} = \left(\mathbf{a}_i \cdot \mathbf{b}_j \right).$$

In particular the product is defined only if the length of a row vector of $\mathbf{A}$ is the same as the length of a column vector of $\mathbf{B}$, i.e., only if $q = r$. Hence, the product is defined only if the number of columns of $\mathbf{A}$ is equal to the number of rows of $\mathbf{B}$.

In our example, try $\mathbf{B} \cdot \mathbf{A}$ in MATLAB, and see what happens.

It is clear from all of the above that the order of the factors in a product is important. In fact, for both $\mathbf{A} \cdot \mathbf{B}$ and $\mathbf{B} \cdot \mathbf{A}$ to be defined, we must have both $q = r$ and $p = s$. This is true for our example vectors $\mathbf{u}$ and $\mathbf{v}$. In MATLAB try both u*v and v*u and see what you get. Be sure you understand the result.

If $p = q = r = s$, then both $\mathbf{A} \cdot \mathbf{B}$ and $\mathbf{B} \cdot \mathbf{A}$ are defined, and both have the same size. In general, however, the two products are not equal. That is, matrix multiplication is *noncommutative* in general. We will explore this further in the exercises.

Matrices and systems of linear equations

One of the most important applications of matrices is to linear systems of algebraic equations. For example, consider the system

$$2x + 4y - 3z = 8$$
$$-2x + 3y + 2z = -6 \tag{1}$$
$$-2x + 4y + z = -4.$$

We can write this system as a matrix equation by introducing the *coefficient matrix*

$$\mathbf{A} = \begin{pmatrix} 2 & 4 & -3 \\ -2 & 3 & 2 \\ -2 & 4 & 1 \end{pmatrix},$$

the *vector of unknowns*

$$\mathbf{u} = \begin{pmatrix} x \\ y \\ z \end{pmatrix},$$

and the *vector of right hand sides*

$$\mathbf{b} = \begin{pmatrix} 8 \\ -6 \\ -4 \end{pmatrix}.$$

With these definitions, the system in (1) can be expressed succinctly as

$$\mathbf{Au} = \mathbf{b}. \tag{2}$$

The matrix equation in (2) is so simple, one is naively led to explore the possibility of solving it by dividing by the matrix $\mathbf{A}$. Remembering that matrix multiplication is not commutative, we realize that we should divide on the left, i.e. since A*u=b, we might expect that u=A\b. In fact, this often works. Let's try it in this case.

```
>> A=[2 4 -3;-2 3 2 ; -2 4 1];
>> b= [8 -6 -4]';
>> u=A\b

u =

     1
     0
    -2
```

We can check to see that this is actually a solution:

```
>> A*u

ans =

        8
       -6
       -4
```

Since this is equal to b, we have a solution.

Once again we have demonstrated the ease with which MATLAB can solve mathematical problems. Indeed the method of the previous paragraph works in great generality. **However, it does not always work!** In order for it to work, it is necessary that the coefficient matrix be *nonsingular* (The term nonsingular will be defined in Chapter 8). Unfortunately, many of the applications of linear systems to ordinary differential equations have coefficient matrices which are singular.

Furthermore, since matrix multiplication is a rather complicated procedure, one can expect that matrix division is at least equally complicated. The above naive approach is hiding a lot of mathematics.

As a result, we will have to study the problem quite thoroughly. We will return to this in Chapter 8.

Array operations

While matrix addition operates element by element, the definition of matrix multiplication is much more complicated. There is often a need for other operations on matrices which operate element by element. In MATLAB these operations are built-in and very easy to use. For example, if we introduce another vector **w** of the same size as **v** and then enter v.*w we see

```
>> w=[9;0;2;3]';
>> v.*w

ans =

        9
        0
       18
       15
```

If you look closely you will see that the result is a vector of the same size as **v** or **w**, and that each element in the result is the product of the corresponding elements in **v** and **w**. The MATLAB symbol for this operation is `.*`, and as you have seen, the result is quite different from what `*` does. Try entering `v*w` and see what happens.

The operation we have just defined is called *array multiplication*. There are other array operations. All of them act element-by-element. Try `v./w` and `w./v`. This is *array right division*. Then try `v.\w` and `w.\v`, and compare the results. This is called *array left division*.

For all array operations it is required that the matrices be of exactly the same size. You might try `A.*v` to see what happens when they are not.

There is one other array operation — *array exponentiation*. As might be expected, this, too, is an element-by-element operation. The operation `A.^4` results in every element of the matrix **A** being raised to the fourth power. For example, with the matrix **A** which we used in the previous section:

```
>> A.^2


ans =

     4    16     9
     4     9     4
     4    16     1
```

If you look closely, you will notice that each element of the result is the square of the corresponding element of **A**. Try `A^2`. Since this is the same as `A*A`, it is quite different from `A.^2`, which is the same as `A.*A`.

The built-in MATLAB functions, which we discussed briefly in Chapter 1, are all designed to be *array smart*. This means that if you apply them to a matrix, the result will be the matrix obtained by applying the function to each individual element. For example:

```
>> cos(A)

ans =

   -0.4161   -0.6536   -0.9900
   -0.4161   -0.9900   -0.4161
   -0.4161   -0.6536    0.5403

>> cos(v)
```

37

```
ans =

    0.5403
   -0.6536
   -0.9972
```

This is an extremely important feature of MATLAB, as you will discover in the next section.

Plotting in MATLAB

None of these array operations would be important if it were not so easy to create and use vectors and matrices in MATLAB. Here is a typical situation. Suppose we want to define a vector that contains a large number $N + 1$ of equally spaced points in an interval $[a, b]$. We can do this quite easily in MATLAB. If $h = (b - a)/N$, we enter t=a:h:b;. For example, with $a = 0$, $b = 1$, and $N = 10$, we have $h = 0.1$. Then using MATLAB we get:

```
>> t=0:0.1:1;
>> t

t =

  Columns 1 through 7

        0    0.1000    0.2000    0.3000    0.4000    0.5000    0.6000

  Columns 8 through 11

   0.7000    0.8000    0.9000    1.0000
```

Thus t is a row vector that contains the desired entries.

Now if we enter y=t.^3, we will get another vector with 11 entries, and each entry in y would be the cube of the corresponding entry in t. We can get a rudimentary plot of the function t^3 by plotting the entries of y versus the entries of t. MATLAB will do this for us.

```
>> y=t.^3;
>> plot(t,y)
```

The result will be a new Figure Window, with the 11 points plotted and connected by straight lines, as indicated in Figure 3.1.

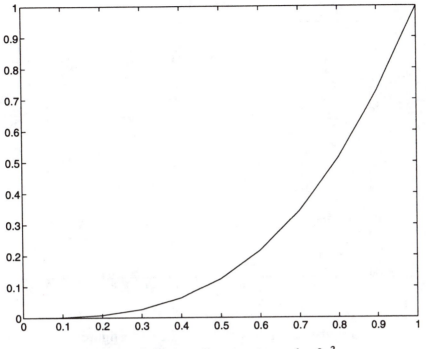

Figure 3.1. A rudimentary graph of t^3.

For a variant of this, try `plot(t,y,'--')`. This results in a dashed curve for the graph. Other *line types* are available — try this command with `'--'` replaced by `'-'`, `':'`, or `'-.'`. If you want to see only the points and not the connecting straight lines, you can plot using a *point type*. Try `'o'`, `'.'`, `'x'`, and `'+'`. If you have a color monitor, it is also possible to choose a variety of colors. See the MATLAB *User's Guide* for instructions.

Frequently it is desirable to plot two graphs on the same figure. This is also easily accomplished in MATLAB. Suppose we want the graphs of $y = x^2 - 3x + 5$ and $z = x^3 + 6x^2 - 6$ over the interval $[-2, 3]$ on the same figure, and that we want the first to be a solid curve and the second a dashed curve. Here is how to do it:

```
>> x=-2:0.05:3;
>> plot(x,x.*x-3*x+5,x,x.^3+6*x.^2-6,'--')
>> grid
```

The command `grid` does just what it says — it adds a grid to the figure. The important command to understand is the `plot` command. If we wanted to make the two plots separately, we would have entered:

```
>> x=-2:0.05:3;
>> plot(x,x.*x-3*x+5)
>> plot(x,x.^3+6*x.^2-6,'--')
```

39

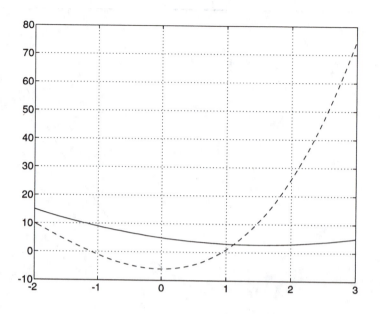

Figure 3.2. Two graphs on the same figure.

This would have the effect of erasing the first plot when the second is made. By putting the two together in the command `plot(x,x.*x-3*x+5,x,x.^3+6*x.^2-6,'--')` we get both on the same figure. The result is shown in Figure 3.2.

An alternative to the procedure in the previous paragraph involves the commands `hold on` and `hold off`. The command `hold on` tells MATLAB to add subsequent plots to the existing figure, without erasing what is already there. The command `hold off` tells MATLAB to return to the standard procedure of erasing everything before the next plot. This means that `hold on` is in effect until a `hold off` command is executed. Thus, we could have used the following sequence of commands:

```
>> x=-2:0.05:3;
>> plot(x,x.*x-3*x+5)
>> hold on
>> plot(x,x.^3+6*x.^2-6,'--')
>> hold off
```

More sophisticated plots are just as easy with MATLAB. For example, a *loglog* plot can be drawn using the command `loglog(x,y)` instead of `plot(x,y)`. Of course, in this case the vectors should have only positive values, since we are really plotting the logarithms of the vectors against each other. Try this:

```
>> x=1:.01:100;
>> y=x.^3+x.^2;
>> loglog(x,y)
>> grid
```

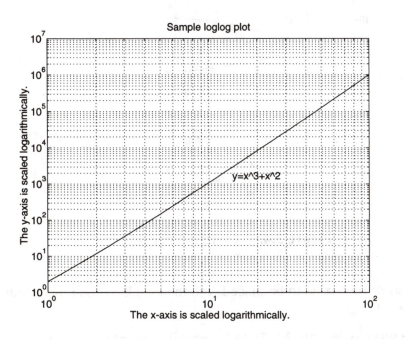

Figure 3.3. A loglog graph.

Other possibilities are `semilogx`, which uses a logarithmic scale for the x-axis and a linear scale for the y-axis. Similarly `semilogy` uses a logarithmic scale for the y-axis and a linear scale for the x-axis.

Plots look better, and they are more informative, if they are properly labeled. This is easy to do. In the previous example, the following commands add a title on top, axis labels, and a label for the graph itself.

```
>> title('Sample loglog plot')
>> xlabel('The x-axis is scaled logarithmically.')
>> ylabel('The y-axis is scaled logarithmically.')
>> gtext('y=x^3+x^2')
```

The last command, `gtext('y=x^3+x^2')`, needs additional action on the part of the user. This command enables the user to place text anywhere on the Figure Window. It is necessary to complete the action by using the mouse to select the point on the Figure Window at which

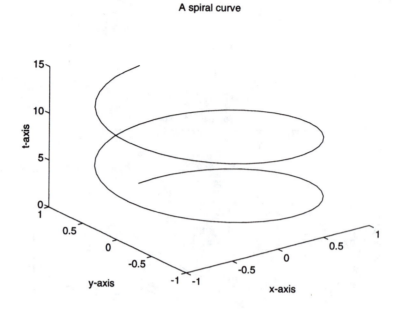

A spiral curve

Figure 3.4. A curve in three dimensions.

the lower left part of the text is to be placed, and then clicking the mouse button. The result is shown in Figure 3.3.

Finally, we want to describe MATLAB's capability for plotting curves in three dimensions. This uses the command plot3. It is used in much the same way that plot is used. For example, if we want to plot the spiral defined by

$$x = \sin(t)$$
$$y = \cos(t)$$
$$z = t$$

for $0 \le t \le 4\pi$, we proceed as follows:

```
>> t=0:0.05*pi:4*pi;
>> plot3(sin(t),cos(t),t)
>> xlabel('x-axis')
>> ylabel('y-axis')
>> zlabel('t-axis')
>> title('A spiral curve')
```

The results are shown in Figure 3.4.

After learning how to plot graphs, the user will want to print them out. Nothing could be easier in MATLAB. To print what appears in the current Figure Window, simply enter print at the MATLAB prompt.

42

Simple function M-files

We have seen that MATLAB has a large variety of built-in mathematical functions. In addition, it is very easy for the user to add to the list. For this purpose, it is necessary to know how to use an editor to create and edit text files on your computer. While the Macintosh version of MATLAB has a built-in editor, the other versions do not. However, there are a wide variety of editors available, and any of these will do. It is even possible to use a word processor, but if you do it is absolutely essential that you save the file as a text file.

As an example of how this works, suppose you will need to frequently compute the function $f(x) = x^3 \cos(x) - 4$. Typing `x^3 * cos(x) -4` at the MATLAB prompt for a large collection of different values of x can get tedious. Instead, let's create a *function M-file* to do this for us. First of all, we need a name for this function. Let's call it funn, so that $funn(x) = x^3 \cos(x) - 4$. Open your editor and create a new file containing the following three lines (one of them is blank, and is provided for readability):

```
function y = funn(x)

y = x^3 * cos(x) -4;
```

Then save the file as `funn.m`.

That's all there is to it. Now if you want to compute funn(0.3) using MATLAB, simply enter `funn(0.3)` at the MATLAB prompt. Try it.

There is one important enhancement you will want to make to the M-file `funn.m`. If you try to compute funn on a matrix or a vector, you will find that it is not array smart. It is always a good idea to make your function M-files array smart, and it is very easy to do so. It is only necessary to replace arithmetic operations with their array smart versions, and that is a matter of adding a period in front of them. For example, the array smart version of funn is:

```
function y = funn(x)

y = x.^3 .* cos(x) -4;
```

There is one other feature of MATLAB that is illustrated in the file `funn.m`. Notice that if x is a matrix, then so is `x.^3.*cos(x)`. On the other hand, 4 is a number, so the difference of these two is not defined in ordinary matrix arithmetic. However, MATLAB allows it, and in cases like this will subtract 4 from every element of the matrix. This is a very useful feature in MATLAB.

Now it is extremely easy to see the graph of funn. The commands

```
>> x=-5:0.05:5;
>> plot(x,funn(x))
```

43

```
>> grid
>> title('y = funn(x)')
```

will result in the graph of funn over the interval $[-5, 5]$. (See Figure 3.5.)

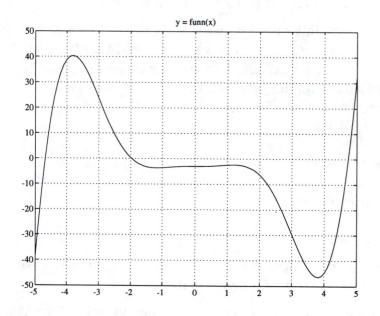

Figure 3.5. Graph of $y = x^3 \cos(x) - 4$.

Another useful thing that MATLAB can do is find the zeros of functions. For example, from Figure 3.5, it will be seen that funn has a zero near $x = 5$. To find out more precisely where this zero is we can use the MATLAB routine `fzero`. We need to tell it the name of a function M-file describing our function and the approximate location of the zero. The syntax in this case is `fzero('funn',5)`. Notice that the function name must be surrounded by single quotes, and that the `.m` part is not needed.

```
>> fzero('funn',5)

ans =

    4.7497
```

You can check the accuracy of this answer.

```
>> funn(ans)

ans =

    3.1086e-14
```

Since this is MATLAB's notation for 3.1086×10^{-14}, we see that our answer is very close to a zero of funn.

As another example, consider the function of two variables $y = x^2 - t$. We need to choose a name for this function, so let's use the mnemonic xsqmt. Then the following M-file describes this function:

```
function y = xsqmt(t,x)

y = x.^2 - t;
```

This file must be saved as xsqmt.m, and then the function can be evaluated at $t = 2$ and $x = 5$ as follows:

```
>> xsqmt(2,5)

ans =

    23
```

Several key facts about function M-files should be pointed out.

- The first line of a function M-file must conform to the indicated format. The very first word must be the word function. The rest of the first line has the form

```
dependent_variable = function_name(independent_variables)
```

In the example xsqmt used above, y=xsqmt(t,x) indicates that t and x are the independent variables and y is the dependent variable.

- The rest of the function M-file defines the function using the same syntax as we have been using at the MATLAB prompt. Remember to put a semicolon at the end of lines in which computations are done. Otherwise the results will be printed in the Command Window.

45

- The name of an M-file is of the form `function_name.m`, where `function_name` is the name you choose to call the function. While this name can be almost anything, there are a few rules:

  - It must start with a letter (either upper case or lower case).

  - The name must consist entirely of letters, numerals, and underscores (_). No other symbols are allowed. In particular, no periods are allowed.

  - The name can be arbitrarily long, but MATLAB will only remember the first nineteen symbols.

  - Do not use names already in use such as `cos`, `plot`, or `dfield`. If you do MATLAB will not complain, but you will not be able to use these names for their original purpose until you erase your own files with these names.

- The variable names used in a function M-file must satisfy the same rules that apply to names of M-files, and which are listed in the previous bullet. Other than that they are arbitrary. As a result the file `funn.m` could have been written as

```
function stink = funn(skunk)

stink = skunk.^3 .* cos(skunk) - 4;
```

  and the results would have been exactly the same. It is important to realize that the variable names in function M-files are *local* to these files; i.e., they are not recognized outside of the M-files themselves.

- Everything on a line in an M-file after a percentage sign (%) is ignored by MATLAB. This can be utilized to put comments in a file. For example, the function M-file `funn.m` could be written as

```
function y = funn(x)

% This is used in homework # 3.
y = x.^3 .* cos(x) - 4;
```

The comment line can and should be used to document your M-files. It is amazing how quickly we forget why we did things the way we did. Comment lines immediately after the function statement can be read using the MATLAB `help` command. For example, if `funn.m` is as above then we get

```
>> help funn

This is used in homework #3.
```

If the `help` command does not provide enough information about a function M-file, the command `type` can be used to print the entire file to the Command Window.

```
>> type funn

function y = funn(x)

% This is used in homework #3.
y=x.^3 .* cos(x) -4;
```

The commands `help` and `type` work with any MATLAB function. Try them on `cos`, or `dfield` for example. Of course, if the file is very long, or if the command is a MATLAB built-in function, `type` may not be too useful. It can be very useful, however, to read the relatively small M-files that you will need to write for the exercises in this Manual.

The use of simple function M-files as indicated in this section greatly enhances the capabilities of MATLAB. However, they represent only the beginning of the programming capabilities available in MATLAB. The routine `dfield` is also an example of a function M-file.

Exercises

1. Consider the matrices

$$A = \begin{pmatrix} 1 & 7 \\ 0 & -3 \end{pmatrix}, \quad B = \begin{pmatrix} -3 & 2 \\ 3 & -2 \end{pmatrix}, \quad \text{and} \quad C = \begin{pmatrix} 3 & 4 & 1 \\ -2 & -1 & 4 \\ 0 & -4 & 3 \end{pmatrix},$$

and the vectors

$$v = \begin{pmatrix} 3 \\ 5 \end{pmatrix}, \quad w = \begin{pmatrix} 2 \\ -9 \end{pmatrix}, \quad x = (-4 \quad 3), \quad \text{and} \quad y = (0 \quad 6 \quad -3).$$

Try the following combinations in MATLAB. Which are defined and which are not? Where a combination is not defined, explain why.

A*A	A*B	A*C	A.*A	A+B	A+C	A./C	A.\B
A*x	x*A	v*A	x.*A	A.*x	y*C	C*x	C*y
A*v	A*w	x+v	v+w	x*x	v.*w	w.*v	y.^2
y^2	A^2	x+y	v*y	v.*y	v*w	x*w	x.*w

2. The matrix

$$I = \begin{pmatrix} 1 & 0 \\ 0 & 1 \end{pmatrix}$$

is called the (2×2) identity matrix. Show that $I \cdot A = A$ and $A \cdot I = A$ for every 2×2 matrix A. Thus I commutes with every 2×2 matrix.

3. Show that the matrices $\mathbf{A} = \begin{pmatrix} 1 & -2 \\ 0 & 3 \end{pmatrix}$ and $\mathbf{B} = \begin{pmatrix} -2 & -3 \\ 0 & 1 \end{pmatrix}$ commute.

4. The MATLAB command `A=rand(2)` will define A as a 2×2 matrix with entries which are chosen randomly from the interval $(0, 1)$. Generate random 2×2 matrices A and B and compare A*B and B*A. Do this ten times and report how many times A*B = B*A. This can be done easily in MATLAB by using the single composite command

```
A=rand(2);B=rand(2);A*B-B*A
```

After executing it once, it can be repeated by using the up arrow to bring it back to the command line. Compare your results with other users.

5. What happens in the previous exercise if you use .* instead of *?

6. Why is there no .+ command in MATLAB?

7. On the same figure plot $y = \cos(x)$ and $z = \sin(x)$ over the interval $[0, 4\pi]$. Use different line styles for each curve, and label the figure appropriately.

8. On the same figure plot the three curves $y = \sin(x)$, $y = x - x^3/6$, and $y = x - x^3/6 + x^5/120$ over the interval $[-3, 3]$. Use different line styles for each curve, and label the figure appropriately. Do you recognize the relationship between these three functions?

9. On the same figure plot the graphs of the function $y = e^x$ and its Taylor approximations of order 1, 2, and 3 over the interval $[-3, 3]$. Use different line styles for each curve, and label the figure appropriately.

10. Consider the functions $y_1 = x$, $y_2 = x^2$, and $y_3 = x^4$ on the interval $[0.1, 10]$. Plot these three functions on the same figure using the command `plot`. Now do the same thing with the other plotting commands `semilogx`, `semilogy`, and `loglog`. Turn in only the one that you think is most revealing about the relationship between these functions. Use different line styles for each curve, and label the figure appropriately. (Plotting more than one curve on a figure using any of these commands follows the same procedure used with `plot`.)

11. In three dimensions plot the curve defined by

$$x = t\cos(t)$$

$$y = t\sin(t)$$

$$z = t$$

over the interval $t \in [0, 4\pi]$. Label the figure appropriately.

12. Write a function M-file to calculate the function $f(x) = \sin(x^2) - 2\cos(x)$. Use it to graph f over the interval $(0, 5)$.

13. Find all of the zeros in the interval $(0, 5)$ of the function f defined in the previous exercise.

14. Write a function M-file to calculate the function $g(t, x) = e^{-t}x^2$. Use it to find an approximate numerical value for g at the points $(0, 1)$, $(1, 2)$, $(-1, 1)$, and $(-2, 3)$.

15. Write a function M-file to calculate $f(z) = \sqrt{|z|}$. Use it to graph f over the interval $(-3, 2)$.

16. Write a function M-file to compute the function $f(t) = t^{1/3}$. Use it to graph the function over the interval $[-1, 1]$. (This is not as easy as it looks. Read the section on complex arithmetic in Chapter 1.)

17. Write a function M-file to calculate the function $f(t) = e^{-t}(t^2 + 4e^t - 5)$. Use it to graph f over the interval $(-2.5, 3)$.

18. Find all of the zeros in the interval $(-2.5, 3)$ of the function f defined in the previous exercise.

19. For the function funn defined in this chapter find all solutions to the equation $funn(x) = 5$ in the interval $[-5, 5]$.

20. Plot the function $f(t) = 5$ over the interval $[0, 2]$. **Remark:** You will find this exercise more difficult than it looks. It is harder to plot constant functions in MATLAB than most nonconstant functions. There are a variety of tricks that will work, but the methods that are most consistent with the ways used for nonconstant functions (and therefore are most useful in programming applications) use the MATLAB commands `size` and `ones`. Use `help` on these commands.

4.   The Symbolic Toolbox

The Symbolic Toolbox provides a link between MATLAB and the symbolic algebra program Maple. Unlike the rest of MATLAB, which provides powerful numerical routines, the Symbolic Toolbox deals with symbols, formulas, and equations. In dealing with differential equations, it leads to explicit formulas for the solutions, provided such formulas exist at all.

The Symbolic Toolbox is a part of version 4 of the Student Edition of MATLAB, however, it is distributed as an addition to standard MATLAB. Therefore it may not be available with your copy. If you are unsure whether or not it is available, execute the command `help symbolic`. If it is available, you will get a list of all of the commands available in the toolbox. If it is not, you will get an error message, saying `symbolic not found`.

For a general introduction to the Symbolic Toolbox find a copy of the *Symbolic Math Toolbox User's Guide*. This provides a very good tutorial on the use of the Symbolic Toolbox. In some ways it provides a much more comprehensive treatment than we will present here.

In this chapter, we will put our emphasis on using the Symbolic Toolbox to solve differential equations. This means that MATLAB will try to tell you the exact, analytic solution to an equation, or even to an initial value problem. MATLABcan be successful, of course, only if such a solution exists.

The Symbolic Toolbox has many other capabilities. It is a very useful program, but it takes some practice to become an expert in its use. You will be impressed with the speed with which MATLAB will solve simple ODEs. On the other hand you will find that the Symbolic Toolbox has some quirks — enough quirks so that it is a good idea to be wary of what it tells you.

Symbolic expressions and basic calculus operations

A *symbolic expression* is a mathematical expression that involves functions, variables, parameters, numbers, equations, etc. Examples are $\cos(x)$, $\sin(x^2)$, $1 - x^2/2$, and $x^2 + y^2 = 25$. MATLAB needs a way to recognize symbolic expressions when they are entered at the command line. The convention is that symbolic expressions must be enclosed between single quotes.

The command `f='sin(x^2)'` will cause MATLAB to recognize that the symbol `f` represents the symbolic expression $\sin(x^2)$ from that point on.

The Symbolic Toolbox contains a very useful and easy to use plotting tool. The command `ezplot(f)` will cause $\sin(x^2)$ to be plotted over the interval $[-2\pi, 2\pi]$. In general the command `ezplot(g)`, where g is the symbolic expression of a function of one variable, will plot the function over a subinterval of $[-2\pi, 2\pi]$. The routine automatically eliminates portions of this interval near the endpoints in which the function is not varying very fast. For example `ezplot('exp(-x^2)')` will result in the graph of e^{-x^2} over an interval which is approximately equal to $[-2.5, 2.5]$. Since e^{-x^2} is very close to 0 if $|x| > 2.5$, ezplot catches the most important part of the graph, and ignores the rest.

The automatic interval selection feature of ezplot is a default which can easily be over-ridden. For example, the command ezplot('exp(-x^2)',[-4,3]) will plot e^{-x^2} over the interval $[-4, 3]$.

MATLAB will do simple calculus. To find the derivative of $\sin(x^2)$, enter the command diff('sin(x^2)'). If f='sin(x^2)' has been previously executed, it is only necessary to enter diff(f). To see what happens when you forget to use the quotes, enter diff(cos(x)).

Higher order derivatives provide no more challenge to the user, although the computer might take a little longer. The command diff(f,4) will calculate the fourth derivative of $\sin(x^2)$, and this illustrates the syntax that is used for higher order derivatives.

The command int can be used to calculate the indefinite integral of a function. For example, having already entered f='sin(x^2)', we can first differentiate f and then integrate:

```
>> f1 = diff(f)

f1 =

2*cos(x^2)*x

>> g = int(f1)

g =

sin(x^2)
```

Thus we have verified the Fundamental Theorem of Calculus in this case.

It should be pointed out that int gives an indefinite integral of its argument, but it does not give the most general formula, which would involve an arbitrary additive constant. This can lead to seeming contradictions. For example:

```
>> int(diff('sin(x)^2'))

ans =

-cos(x)^2
```

This may seem to contradict the Fundamental Theorem of Calculus. However, the Fundamental Theorem says that $\int f'(x)\,dx = f(x) + C$, where C is an arbitrary constant. Since $\sin(x)^2 = -\cos(x)^2 + 1$, there is no contradiction here. This does illustrate the fact that because it chooses a specific constant of integration, `int` can be somewhat capricious in its choice of an indefinite integral.

We can also use `int` to evaluate definite integrals. For example, to evaluate $\int_0^\pi \sin(x)\,dx$, we would enter `int('sin(x)','0','pi')`, and get 2 for an answer. The endpoints do not have to be numbers. We can just as easily use symbols for the end points. To calculate $\int_a^b \sin(x)\,dx$ we enter:

```
>> int('sin(x)','a','b')

ans =

-cos(b)+cos(a)
```

The Symbolic Toolbox will sometimes come up with answers which have a very strange and complicated appearance. As an example, we will verify the Fundamental Theorem for the function $f(x) = \sin(x)/(x + \cos(x))$.

```
>> f='sin(x)/(x+cos(x))';
>> f1 = diff(f);
>> g = int(f1)

g =

(2*tan(1/2*x)+4*tan(1/2*x)^3+2*tan(1/2*x)^5)/(1+tan(1/2*x)^2)^2
        /(x+x*tan(1/2*x)^2+1-tan(1/2*x)^2)
```

This time our answer looks nothing like f. In fact the expression is so complicated it is difficult to figure out what it even says. There is a nice command in the Symbolic Toolbox which helps in times like this. The command `pretty` will cause the expression to be displayed in a form which is much easier to read than the usual MATLAB display. For this example, execute `pretty(g)`. The result can be recognized as

$$g(x) = \frac{2 \tan(\frac{x}{2})^5 + 2 \tan(\frac{x}{2}) + 4 \tan(\frac{x}{2})^3}{\left(1 + \tan(\frac{x}{2})^2\right)^2 \left(x + x \tan(\frac{x}{2})^2 + 1 - \tan(\frac{x}{2})^2\right)}.$$

While we can understand what the formula says, it is still nothing like f. This is a frequent occurrence when we use symbolic algebra programs. Has MATLAB made a mistake? No, but its integration routine caused it to look at the problem in an unusual way. There is a convenient method to resolve the issue using the command `simple`, which tries to algebraically simplify

52

its argument:

```
>> h = simple(g)

h =

sin(x)/(x+cos(x))
```

This is equal to the function f we started with. The conclusion is that the expression for g above is actually equal to f, although it is not obvious. The command `simple` causes MATLAB to try several different ways of simplifying its argument, and then it reports the shortest one. To see the results of the various attempts, enter `simple(g)` without assigning a name as we did before. Try it. You might be surprised.

There are a variety of commands in MATLAB that can be used to simplify expressions. Included are `simplify`, `expand`, `factor`, and `collect`. `simple` uses all of these, and more, and it will usually give an answer that is good enough. If the occasion arises where you need more, or if you are simply curious (good for you!), use the `help` command, or consult the Symbolic Toolbox *User's Guide*.

Verifying solutions to differential equations

Suppose you are given the differential equation $y' + 2ty = 0$, and the function $y(t) = e^{-t^2}$. It is easy, in this case, to verify directly that y is a solution by substituting y into the differential equation. In more difficult situations it will be helpful to let MATLAB do the hard algebra.

We know how to differentiate y in MATLAB, but in order to verify that y is a solution to the above equation, we must multiply $2t$ times y and then add the result to the derivative of y. It turns out that such elementary arithmetic operations are not as simple in MATLAB as we might expect. For example, let's try what might seem to be the obvious thing

```
>> y = 'exp(-t^2)';
>> diff(y)+'2*t*y'
??? Error using ==> +
Matrix dimensions must agree.
```

Clearly that does not work. (And we learn the surprising fact that MATLAB thinks these expressions are matrices!)

The problem lies in the use of `+`. This symbol cannot be used in MATLAB to add symbolic expressions. Instead MATLAB has special commands for the four arithmetic operations when they are applied to symbolic expressions. They are `symadd`, `symsub`, `symmul`, and `symdiv` for addition, subtraction, multiplication, and division. The use of each of these is essentially the same. For example to multiply $2t$ times y, we use `symmul('2*t',y)`. Notice that we needed quotes around 2*t, since we must make that into a symbolic expression, but we did not need them around y since that has already been defined as a symbolic expression. We get

```
>> exp1 = symmul('2*t',y)

exp1 =

2*t*exp(-x^2)
```

Try `symsub('2*t',y)`, `symadd('2*t',y)`, and `symdiv('2*t',y)`.

Having `exp1` available, we can easily check if y satisfies the differential equation, since this is the sum of `diff(y)` and `exp1`.

```
symadd(diff(y),exp1)

ans =

0
```

The 0 which results ensures us that $y' + 2ty = 0$.

Of course, we can also verify that y is a solution to the differential equation in one fell swoop:

```
>> symadd(diff(y),symmul('2*t',y))

ans =

0
```

There is yet a third way, which is probably the easiest of all. This uses the command `symop`. While the commands `symadd`, etc. only accept two arguments, `symop` accepts up to 16. It takes the arguments, concatenates them, and then simplifies the resulting expression. The easiest way to explain how it works is to give an example. In the previous case we would enter:

```
>> symop(diff(y),'+','2*t*',y)

ans =

0
```

Notice that we had to use `symop` with 4 arguments, each of which is a symbolic expression. The symbolic expressions, when put together, form the differential equation. Notice also that

54

we had to isolate the operation +, but we did not have to isolate *. We also had to isolate diff(y) and y. These do not have to be delimited by single quotes, since they are names of symbolic expressions.

As another example, consider the differential equation $y'' - 2y' + 5y = 0$. We want to verify that $y(t) = e^t \sin(2t)$ is a solution.

```
>> y = 'exp(t)*sin(2*t)';
>> symop(diff(y,2),'-','2*',diff(y),'+','5*',y)

ans =

0
```

Again we broke the differential equation into a number of symbolic expressions. In doing so we separated out the expressions involving y and its derivative, and the operations + and -. This is the procedure to use in general.

The choice of variables

The reader might have noticed that when we used the command diff we did not tell MATLAB which variable to use as the independent variable. Frequently, and in all cases examined so far, there is only one possible choice. However, suppose we asked MATLAB to differentiate or integrate the expression $t \log s$. If you try it, you will find out that MATLAB chooses to differentiate with respect to t. Unless given specific directions, MATLAB will choose a default symbolic variable according to what is called the "symvar rule:"

> The default symbolic character in a symbolic expression is the single, lower case letter, other than 'i' and 'j', that is closest to 'x' in the alphabet. If there are two equally close, the one which is later in the alphabet is chosen.

If you are curious about what the default variable is in a symbolic expression, you can find out by using the command symvar. For example, symvar('w*y') returns y, and symvar('w*z') returns w.

Thus diff('x*y') will differentiate the expression with respect to x. But what if we really want to differentiate with respect to the variable y? To accomplish this, we enter the independent variable as an optional additional argument:

```
>> diff('x*y','y')

ans =

x
```

What do you think happens if we differentiate $f(x, y) = xy$ with respect to z? Try `diff('x*y','z')`.

To take two or more derivatives with respect to y, we use an additional parameter. For example, `diff('x*y^2','y',2)` will result with the second derivative of xy^2 with respect to y, and `diff('cos(s+2*t)',3)` with the third derivative of $\cos(s + 2t)$ with respect to t. The additional arguments must be entered in the correct order. For example `diff('x*y^2',2,'y')` will result in an error message.

The designation of an independent variable in this way can be used with many of the commands in the Symbolic Toolbox, including `int`. The `symvar` rule results in x being chosen as the default variable when no other choices are present in the symbolic expression. Often in differential equations we will want t to be the independent variable, so we will have to specify it in our commands.

Simple first order differential equations

The most important command in the Symbolic Toolbox for solving differential equations is the routine `dsolve`. We will describe the use of that command for first order equations in some detail in this chapter. To get a quick description enter `help dsolve`. This will bring a lot of information about `dsolve` to the command window. On most computers there will be more information than will fit on the screen, and most of it will be lost. To prevent this, you will find it useful to enter `more on` before executing the `help` command. After this the information will be presented one screenful at a time. Be sure to execute `more off` after you are through using `help`.

Perhaps the best way for us to start is with an example. If we want to find the general solution to the equation

$$\frac{dy}{dt} + y = te^t,$$

we enter `dsolve('Dy+y=t*exp(t)')`. The result of the calculation as seen on the computer screen is the following:

```
>> dsolve('Dy+y=t*exp(t)')

ans =

1/2*t*exp(t)-1/4*exp(t)+exp(-t)*C1
```

Now let's look at the syntax we used. The symbol D stands for differentiation in this context, and with this convention `'Dy+y=t*exp(t)'` is a symbolic expression for the differential equation. Notice that the answer is another symbolic expression, which has been assigned to the default MATLAB output variable `ans`. It was not assigned to `y`, as one might expect. To do that we have to enter `y = dsolve('Dy+y=t*exp(t)')`. Try it. Then, to gain a better understanding of how `dsolve` works, try `u = dsolve('Dy+y=t*exp(t)')`.

Let's look at another example. This time we want to solve the initial value problem $y' + y = t = 1$, with y(0)=1. We proceed as follows:

```
>> y=dsolve('Dy+y=t+1','y(0)=1')

y =

t+exp(-t)
```

When we solve an initial value problem, we are really solving a system of equations, so we enter the differential equation and the initial condition as two different symbolic expressions, separated by a comma.

To see what the solution looks like, enter `ezplot(y,[-1,3])`.

Nonlinear equations

In many cases nonlinear equations are no harder for MATLAB and the Symbolic Toolbox than the linear equations solved above. Let's look at the differential equation $2tyy' = 3y^2 - t^2$. This is a more difficult equation than the first one, but MATLAB handles it just as easily. This time, since we will refer to the equation more than once, let's give it a name:

```
>> eq = '2*t*y*Dy = 3*y^2-t^2'

eq =

2*t*y*Dy = 3*y^2-t^2
```

Notice that now the symbol `eq` stands for the entire equation as one symbolic expression. When we want to solve the equation we can use `eq` without the single quotes.

```
>> y = dsolve(eq)

y =

[-t*(1+t*C1)^(1/2)]
[ t*(1+t*C1)^(1/2)]
```

The output is two different solutions. How can this be? If we solve the equation ourselves, we come up with the solution

$$y(t)^2 = t^2 + t^3 C1.$$

This is an implicit formula for the solution of the equation. MATLAB has solved this quadratic equation for *y* and given us both answers.

If we want to look further at the second of these we use the command `sym`:

```
>> y2 = sym(y,2,1)

y2 =

t*(1+t*C1)^(1/2)
```

The syntax requires some explanation. It is first necessary to understand that `y` is a matrix of symbolic expressions. It has 2 rows and 1 column. The command `y2 = sym(y,2,1)` asks MATLAB to assign to the symbol `y2` the expression in the second row and the first column of `y`.

Now let's solve the same equation with a couple of different initial values. First we want the solution with $y(1) = 2$.

```
>> y=dsolve(eq,'y(1)=2')

y =

[-t*(1+3*t)^(1/2)]
[ t*(1+3*t)^(1/2)]
```

In this case it is the second of these which is the correct solution. To isolate it we enter:

```
>> y=sym(y,2,1)

y =

t*(1+3*t)^(1/2)
```

Next we look for the solution with $y(1) = -2$.

```
>> y=dsolve(eq,'y(1)=-2')

y =

[-t*(1+3*t)^(1/2)]
[ t*(1+3*t)^(1/2)]
```

These are the same two formulas we got in the previous case, but now it is the first one which is correct. Hence:

```
>> y=sym(y,1,1)

y =

-t*(1+3*t)^(1/2)
```

In each case MATLAB found two solutions, and in each case only one is correct. In situations like this, it is up to the user to decide which answer is the correct one. As in the case here, it is often only necessary to check the initial value.

What happens if an equation has an implicitly defined solution, which cannot be explicitly represented? Here is one example.

```
>> dsolve('(5*y^4+1)*Dy=sin(t)')

ans =

RootOf(Z^5+Z+cos(t)-C1)
```

In this case the implicit solution is given by $y^5 + y + \cos(t) = C1$. MATLAB discovered that, and it is smart enough to know that there is no explicit formula for the root of this fifth order polynomial equation.

On the other hand, consider the separable equation $(1 - \sin(x))x' = t$, with initial condition $x(0) = 0$. It is easy to see that the solution is defined implicitly by the equation

$$x + \cos(x) = t^2/2 + 1.$$

Nevertheless MATLAB cannot find a solution:

```
>> dsolve('(1-sin(x))*Dx=t','x(0)=0')
??? Error using ==> dsolve
Explicit solution could not be found.
```

Although MATLAB generally does a good job of solving first-order equations, it frequently will not find an implicitly defined solution, unless it has some way of expressing the solution to the implicit equation. It is clever at doing that, however. As an example try to solve the equation $(1 + e^y)y' = \cos(t)$. You can use the help command to help you understand MATLAB's answer.

Solving algebraic equations

The Symbolic Toolbox provides a convenient way to find exact solutions to algebraic equations (i.e. equations which do not involve derivatives). For example, to find the roots of the quadratic polynomial ax^2+bx+c, we can use the command x = $solve('a*x^2+b*x+c=0')$. Follow this by `pretty(x)`, and we rediscover the quadratic formula.

Notice that although there are 4 symbols (i.e `a`, `b`, `c`, and `x`) in the equation, MATLAB solves for x. This is an example of the `symvar` rule in action. Perhaps we really want to solve this equation for a. If so we execute x = $solve('a*x^2+b*x+c=0','a')$.

The `solve` command is quite versatile. It will solve systems of equations, provided explicit solutions exist. For example, let's solve the equations

$$x^2 - 2xy - 3 = 0$$
$$y^2 + 3xy = 0.$$

In MATLAB we first enter the equations, and then use `solve`.

```
>> eq1='x^2-2*y*x-3=0';
>> eq2='y^2+3*x*y=0';
>> solve(eq1,eq2)

ans =

y = 0, x = -3^(1/2)
y = 0, x = 3^(1/2)
x = -1/7*3^(1/2)*7^(1/2), y = 3/7*3^(1/2)*7^(1/2)
x = 1/7*3^(1/2)*7^(1/2), y = -3/7*3^(1/2)*7^(1/2)
```

We get four solution pairs. Actually, the solutions are better presented if we enter `[x,y]` = `solve(eq1,eq2)`. You should try this here, and we will illustrate it in the next paragraph.

Suppose the equations have parameters in them, like a and b in

$$x^2 - 2xy - a = 0$$
$$y^2 + bxy = 0.$$

MATLAB can handle this situation, too, but, when there are parameters, it is necessary to tell MATLAB which of the variables should be expressed in terms of the others.

```
>> eq1='x^2-2*y*x-a=0';
>> eq2='y^2+b*x*y=0';
>> [x,y]=solve(eq1,eq2,'x,y')

x =

[                -a^(1/2)]
[                 a^(1/2)]
[-a^(1/2)/(1+2*b)^(1/2)]
[ a^(1/2)/(1+2*b)^(1/2)]

y =

[                       0]
[                       0]
[ b*a^(1/2)/(1+2*b)^(1/2)]
[-b*a^(1/2)/(1+2*b)^(1/2)]
```

Here we see listed the vectors x and y, each with four entries. Each pair of corresponding entries, e.g. $(-\sqrt{a}, 0)$, is a solution pair for the two equations.

The command `solve` has one more property that is quite useful, and is not usually found in symbolic algebra programs. If MATLAB cannot solve the equations symbolically, it will try to find a numerical solution and, if it is successful, that result will be returned. For example, the numbers x such that $x = \cos(x)$ cannot be expressed in any explicit form, but MATLAB will give a numerical answer:

```
>> solve('x=cos(x)')

ans =

.7390851332151606
```

The logistic equation

The *logistic equation* provides a mathematical model of the growth of various species in non-competitive circumstances. The equation is

$$p' = ap - bp^2.$$

The constants a and b which appear in the equation are different for different species, and $p = p(t)$ represents the number of individuals in the species. Usually we have $a > 0$ and $b \geq 0$.

The rationale behind this model is that for small populations, the $-bp^2$ term is small and negligible with respect to the ap term. Hence the equation is approximated by $p' = ap$, and the solution should be approximately equal to $p_0 e^{at}$, where $p_0 = p(0)$. This agrees with the observation, originally due to Malthus, that in ideal circumstances, populations grow exponentially.

On the other hand, as p increases the $-bp^2$ term is no longer negligible. Now the rate of increase, $ap - bp^2$, decreases, reflecting the fact that the size of the population is limiting the growth of the population. Clearly, $p' < 0$ if $p > a/b$ and $p' > 0$ if $p < a/b$. It is not difficult to see from this that $p \to a/b$ as $t \to \infty$.

This is immediately verified when we use MATLAB to solve the logistic equation.

```
>> leq = 'Dp=a*p-b*p*p'

leq =


Dp=a*p-b*p*p

>> p = dsolve(leq,'t');
>> pretty(p)
```

$$
-\ \frac{a}{-\,b\,-\,\exp(-\,a\,t)\ C1\ a}
$$

Clearly, no matter what the constant C1 is, the exponential term approaches 0 for large values of t, and we see that $p \to a/b$ as $t \to \infty$.

Let's look for the solution with the initial value $p(0) = p_0$. We will use pzero in MATLAB to stand for p_0.

```
>> p = dsolve(leq,'p(0) = pzero','t');
>> pretty(p)
```

$$
-\ \frac{a}{-\,b\,-\,\dfrac{\exp(-\,a\,t)\ (a\,-\,\text{pzero}\ b)}{\text{pzero}}}
$$

Looking at a specific example is revealing and allows us to introduce the command subs, which is used to substitute for symbolic expressions within other symbolic expressions. The syntax is subs(e1,e2,e3), where e1, e2, and e3 are symbolic expressions, and e3 appears within **e1**. The effect is that every occurrence of **e3** in **e1** is replaced by **e2**. We will see how that works in this example.

Suppose we have a population (perhaps of some single celled being) that begins with 10 individuals at time $t = 0$. We discover that the population doubles after one day, and that as time increases the population levels off at 500 individuals. We want to find the specific solution to the logistic equation that models this behavior.

First, since $p(0) = 10$, we want to replace `pzero` by 10:

```
>> p = subs(p,10,'pzero');
>> pretty(p)
```

$$
-\ \frac{a}{-\ b\ -\ 1/10\ \exp(-\ a\ t)\ (a\ -\ 10\ b)}
$$

Next, since the population levels off at 500, we know that $a/b = 500$. To ensure this we substitute $a/500$ for b.

```
>> p = subs(p,'a/500','b')

p =

-a/(-1/500*a-49/500*exp(-a*t)*a)
```

Finally we will use the fact that $p(1) = 20$ to solve for a. First we find the symbolic expression for $p(1)$ by substituting 1 for t.

```
>> p1 = subs(p,1,'t');
>> p1 = simple(p1)

p1 =

500/(1+49*exp(-a))
```

It is easy enough to set this equal to 20, and to solve for a, but let's get MATLAB to do it. First, we have to determine how to write the equation $p(1) = 20$ in MATLAB. Notice that p1 is already a symbolic expression. We add the =20 part to it using `symop`.

```
>> eqa = symop(p1,'=','20')

eqa =

500/(1+49*exp(-a))=20
```

Now we must solve this equation for a, but MATLAB does that easily.

```
>> aa = solve(eqa)

aa =

-log(24/49)
```

At last we are ready to find the final expression for the population by substituting this value for a.

```
>> p = subs(p,aa,'a')

p =

log(24/49)/(1/500*log(24/49)+49/500*exp(log(24/49)*t)*log(24/49))
```

We can view the population graphically by entering `ezplot(p,[0,15])`. Doing this we get a complicated expression along the top, which is not too appealing. We will, therefore, replace it with something more to our taste. In addition we will add some descriptive labels to the axes. The entries in the Command Window which accomplish all of this are:

```
>> ezplot(p,[0,15])
>> title('The growth of a population according to the logistic equation.'
>> xlabel('time')
>> ylabel('Population')
```

The result is shown in Figure 4.1

The formula for the population given above, and the plot in Figure 4.1 give a lot of information about the solution, but suppose we want to know more precisely the value of the population at, say, $t = 5$. It is easy to get a formula for this. We substitute 5 for t in the formula for the solution: `p5 = subs(p,5,'t')`. The result is a rather daunting expression. If we want to know the numerical value of the resulting complicated expression, we can use the command `numeric` which gives the numerical value of a symbolic expression. In this case the command is `numeric(p5)`.

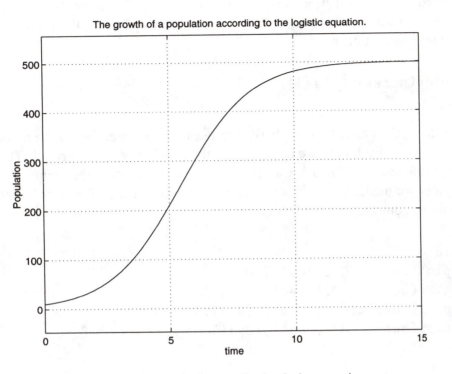

Figure 4.1. A solution to the logistic equation.

Plotting symbolic expressions

The command `ezplot` provides a quick and easy way to plot functions defined by symbolic expressions. However, we will frequently want to develop more elaborate plots, perhaps containing more than one curve. Perhaps the symbolic expression contains a parameter, and we want to plot the function for several values of the parameter. With some simple modifications, `ezplot` will handle these more complicated situations.

We will consider the differential equation

$$x' = -\cos(2\pi t)\, x + \epsilon x,$$

where ϵ is a small constant, and we want to know how the solution with initial value $x(0) = 1$ changes as ϵ changes. The initial value problem is easily solved by MATLAB.

```
>> x = dsolve('Dx=-cos(2*pi*t)*x + ee*x','x(0)=1')

x =

exp((-cos(pi*t)*sin(pi*t)+t*ee*pi)/pi)
```

We have used `ee` as a substitute for ϵ.

65

We want to plot this function for $\epsilon = 0, \pm 0.1, \pm 0.2$. To make the substitution, say of $\epsilon = 0$, we can use `subs(x,0,'ee')`. Then we can use `ezplot` on the result. However, it is easier to do everything at once.

```
>> ezplot(subs(x,0,'ee'),[0,4])
```

This results in the graph for $\epsilon = 0$. We can use the same method for the other values of ϵ, but there is one hitch. Each time we use `ezplot`, the previous graph is erased. The way around this difficulty is the command `hold on`. This retains the contents of the graphics window when subsequent plots are made, and it does this until the command `hold off` is issued. So we can add the remaining four graphs as follows:

```
>> hold on
>> ezplot(subs(x,0.1,'ee'),[0,4])
>> ezplot(subs(x,-0.1,'ee'),[0,4])
>> ezplot(subs(x,-0.2,'ee'),[0,4])
>> ezplot(subs(x,0.2,'ee'),[0,4])
>> hold off
```

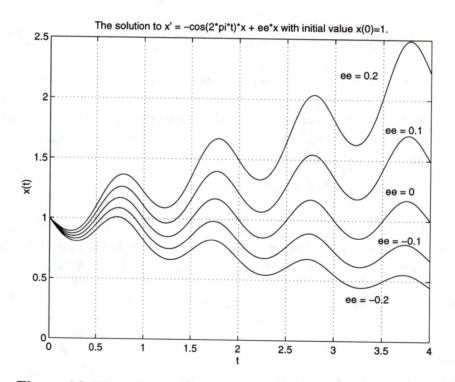

Figure 4.2. The effect of the change of a parameter on a solution.

66

You will notice that the graphics window is rescaled with each new plot. The final picture is less than satisfactory, since much is missing. This can be corrected using the `axis` command, which gives us complete control over the size of the graphics window. The syntax is `axis([xmin,xmax,ymin,ymax])`, where `xmin` and `xmax` are the minimum and maximum values of the coordinate in the horizontal direction, and `ymin` and `ymax` are the minimum and maximum values of the coordinate in the vertical direction. In our case,

```
>> axis([0,4,0,2.5])
```

is a good choice.

Now the graphs look good, but the title on the top is not quite right. To change it, execute

```
>> title('The solution to x'' = -cos(2*pi*t)*x + ee*x with x(0) = 1.')
```

Notice the `x''`. This looks like the second derivative of x, but if you execute the command you will see only one prime in the title of the figure. This is because MATLAB uses the prime to indicate the end of an expression, and it needs another symbol `'` to indicate the prime. Notice also that we used `ee` in the title. There is no easy way to get Greek letters into labels in MATLAB.

Getting back to the project of enhancing the figure, we notice that the horizontal axis has the label t. `ezplot` did this for us, and it is a good choice. The vertical axis is unlabeled, however. Since this axis shows the values of the function $x(t)$, it is natural to label it with $x(t)$.

```
>> ylabel('x(t)')
```

Notice that the command is `ylabel`. If we want to change the label on the horizontal axis, the corresponding command is `xlabel`.

Things are looking better, but it is impossible for an observer to figure out which curve goes with which value of ϵ. We should label each of the curves. To do this we use the command `gtext`, for example `gtext('ee = 0')`. When you execute this, nothing seems to happen. However if you move the cursor into the graphics window, it changes into a large plus sign. Move the cursor to the point where you want to put the lower left corner of the label and click the mouse. The label will appear. Doing the same for the other curves completes our project. The result should look something like Figure 4.2.

It is possible to change the title, xlabel, and ylabel on a figure by reentering the command. The same is not true with text that has been added with `gtext`. In this case it is necessary to remove the old text first. Doing that is a two step process. First, click the mouse somewhere in the middle of the text, and then issue the command `delete(gco)`. The offending text will go away. The same process can be used to erase individual curves or any graphics object. The effect of clicking the mouse is to select the `current object`. Then `gco` is the MATLAB command for get current object.

Exercises

1. Use MATLAB to verify the Fundamental Theorem for the function $f(x) = x \sin^2(x)$. You may have to provide a little algebra of your own.

2. In each of the following cases, use MATLAB and the Symbolic Toolbox to verify that y is a solution to the indicated equation.

   a) $y(t) = 1 + e^{-t^2/2}$, $y' + ty = t$.

   b) $y(t) = 1/(x - 3)$, $y' + y^2 = 0$.

   c) $y(t) = 10 - t^2/2$, $y * y' + ty = 0$.

   d) $y(x) = \log(x)$, $(2e^y - x)y' = 1$.

   e) $y(t) = e^t \cos(2t)$, $y'' - 2y' + 5y = 0$.

3. Use `dsolve` to find the general solution to the following equations.

   a) $y' + ty = t$.

   b) $y' + y^2 = 0$.

   c) $y * y' + ty = 0$.

   d) $(2e^y - x)y' = 1$.

   e) $(x + y^2)y' = y$.

   f) $x(y' - y) = e^x$.

   g) $y'' - 2y' + 5y = 0$.

4. Use `dsolve` to find the solution to the following initial value problems. Plot each solution over the indicated interval(s).

   a) $y' + ty = t$, $y(0) = -1$, $[-4, 4]$.

   b) $y' + y^2 = 0$, $y(0) = 2$, $[0, 5]$.

   c) $y * y' + ty = 0$, $y(1) = 4$, $[-4, 4]$.

   d) $(2e^y - x)y' = 1$, $y(0) = 0$, $[-5, 5]$.

   e) $(x + y^2)y' = y$, $y(0) = 4$, $[-4, 6]$.

   f) $x(y' - y) = e^x$, $y(1) = 4e$, $[0.0001, 0.001]$, and $[0.001, 1]$.

5. Suppose we start with a population of 100 individuals at time $t = 0$, and that the population is correctly modeled by the logistic equation. Suppose that at time $t = 2$ there are 200 individuals in the population, and that the population reaches steady state with a population of 1000. Plot the population over the interval $[0, 20]$. What is the population at time $t = 10$? (Use the command `numeric`. Use `help` on `numeric`, if necessary.)

6. Suppose we have a population which is correctly modeled by the logistic equation, and experimental measurements show that $p(0) = 50$, $p(1) = 150$, and $p(2) = 250$. Use the Symbolic Toolbox to derive the formula for the population as a function of time. Plot the population over the interval $[0, 5]$. What is the limiting value of the population? What is $p(3)$?

7. Suppose we have a population which is modeled by the equation

$$p' = 2p - p^2/1000,$$

where the independent variable is measured in days. What is the limiting value of the population? Consider the solutions with initial populations of 10, 50, 100, 500, and 1000. Plot all five of the

68

solutions on the same figure over a suitable interval of time. For each of these solutions, how long does it take for the population to reach 90% of the limiting population?

8. A certain lake has a volume of V km^3. It is fed by a river at a rate of r_i km^3/year, and there is another river which is fed by the lake at a rate which keeps the volume of the lake constant. In addition there is a factory on the lake which introduces a pollutant into the lake at the rate of p km^3/year. This means that the rate of flow from the lake into the outlet river is $(p + r_i)$ km^3/year. Let $x(t)$ denote the volume of the pollutant in the lake at time t, and let $c(t) = x(t)/V$ denote the concentration of the pollutant.

   a) Show that, under the assumption of immediate and perfect mixing of the pollutant into the lake water, the concentration satisfies the differential equation $c' + ((p + r_i)/V)c = p/V$.

   b) Assuming that the concentration at the initial time $t = 0$ is c_0, derive the formula for the concentration as a function of t.

   c) Show that, whatever the initial concentration, the concentration approaches the limit $p/(p+r_i)$ as $t \to \infty$.

9. In the previous problem suppose that $V = 100$, $r_i = 50$, and $p = 2$. Suppose that the factory starts operating at time $t = 0$, so that the initial concentration is 0.

   a) Derive the equation for the concentration as a function of t.

   b) It has been determined that a concentration of over 2% is hazardous for the fish in the lake. How long will it take until this concentration is reached?

   c) What is the limiting concentration?

   d) Plot the concentration over an appropriate interval.

10. Suppose the factory in the previous problem stops operating at time $t = 0$, and that the concentration was 3.5% at that time. Notice that $p = 0$ in this exercise.

   a) Derive the equation for the concentration as a function of t.

   b) How long will it take before the concentration falls below 2%, and the lake is no longer hazardous for fish?

   c) Plot the concentration over an appropriate interval.

11. Consider the differential equation
$$(1 + t^2)y' = 1 + y^2.$$

   a) Find the general solution.

   b) Find the solution with initial value $y(0) = a$.

   c) On one figure plot the solutions with initial values 0, +1, and -1 over the interval $[0, 3]$. Explain any apparent discontinuities you observe.

5.   Numerical Methods for ODEs

Numerical methods for solving ordinary differential equations are discussed in many textbooks. Here we will discuss how to go about using some of them in MATLAB. In particular we will be interested in doing some experimental error analysis. In line with the philosophy that we are not emphasizing programming in this manual, MATLAB routines for these numerical methods will be made available.

Our discussions of numerical methods will be very brief and incomplete. The assumption is that the reader is using a textbook in which these methods are described in more detail.

Euler's method

We will start with Euler's method. In this very geometric method, we go along the tangent line to the graph of the solution to find the next approximation. I.e., to find an approximation of the solution to the initial value problem $y' = f(t, y)$, with $y(a) = y_0$, on the interval $[a, b]$, we set $t_0 = a$, we choose a step size h, and then we inductively define

$$\left. \begin{aligned} t_{k+1} &= t_k + h \\ y_{k+1} &= y_k + h\,f(t_k, y_k) \end{aligned} \right\} \quad \text{for } k = 0, 1, 2, \ldots, N,$$

where N is large enough that $t_N \geq b$. This algorithm is available in the MATLAB command `eul`.*

To demonstrate how to use `eul`, let us consider the equation $y' = y + t$, with initial condition $y(0) = 1$ on the interval $[0, 3]$. Suppose that you have entered the function $\text{ypt}(t, y) = y + t$ into a function M-file with the name `ypt.m`. Choose the step size $h = 0.3$. Then enter

```
[t,y]=eul('ypt',0,3,1,0.3);
```

The general syntax is

```
[t,y]=eul('dfile',t0,tfinal,y0,stepsize);
```

---

* The function `eul`, as well as `rk2` and `rk4` described later in this chapter, are function M-files which are not distributed with MATLAB. Type `help eul` to see if they are installed correctly on your computer. If not, see the Preface for instructions on how to obtain them. The M-files defining these commands are printed in the appendix to this chapter for the illumination of the reader.

where the differential equation is $y' = \text{dfile}(t, y)$, and the other parameters are fairly obvious. Actually, the step size is an optional parameter. If you were to enter

```
[t,x]=eul('dfile',t0,tfinal,y0);
```

the program would choose a default step size equal to `(tfinal-t0)/100`.

The output consists of two column vectors t and y representing the collection of the t_k's and the corresponding y_k's. To see the approximate solution graphically, it is only necessary to enter `plot(t,y)`. Try it and see.

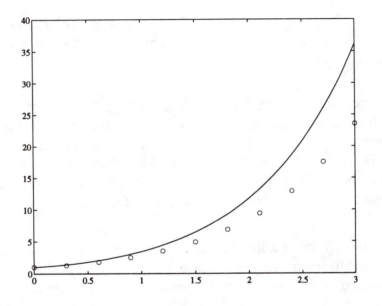

Figure 5.1. The results of Euler's method with step size 0.3

Now we want to examine the error in the numerical solution. One of the reasons that we chose the equation $y' = y + t$ is that we can find the exact solution in order to compare it with the approximate solution. With initial value $y(0) = 1$, the solution is

$$y(t) = -1 - t + 2e^t.$$

We proceed as follows:

```
>> t=0:0.05:3;
>> x=-1-t+2*exp(t);
>> [s,y]=eul('ypt',0,3,1,0.3);
>> plot(t,x,s,y,'o')
```

The first two commands calculate the exact solution. The third is the command that uses Euler's method to compute an approximate solution with step size 0.3. The final command plots both the exact solution with the standard line style, and the approximate solution with the circle point style. The result will be Figure 5.1, which provides a visual image of the accuracy of Euler's method using this step size (not too good in this case).

We would like to issue this same set of commands with different step sizes to analyze graphically how step size affects the accuracy of the approximation. It will quickly get tedious to type these four commands into the Command Window. Instead let's type them once into an editor and save the file with the name `batch.m`, or whatever else you want to call it (as long as it ends in `.m`). To be precise, this file should contain the following four lines:

```
t=0:0.05:3;
x=-1-t+2*exp(t);          % The exact solution.
[s,y]=eul('ypt',0,3,1,h); % The approximate solution.
plot(t,x,s,y,'o')         % Plot both.
```

This file is an example of a *script M-file*. A script M-file is a list of MATLAB commands saved as a file with a name ending in `.m`. Script M-files are very similar to function M-files. In fact, they are simpler. The one obvious difference is that no function statement is needed in a script M-file (as is required in a function M-file). When the name of a script M-file (without the `.m`) is entered at the MATLAB prompt, the list of commands in the file are executed in the order of their appearance. This makes it very easy to execute fairly complicated sequences of commands repeatedly.

Thus, entering `batch` will result in MATLAB executing the above commands. We can add comments as indicated to remind us later what is going on. Notice that we have replaced the step size with the symbol `h`. This symbol will have to be defined before MATLAB will be able to execute the third command in `batch` which has `h` as a parameter. Consequently we must enter, as an example, `h=0.1;batch`. MATLAB will execute the command `h=0.1`, which defines `h`, and then all of the commands in `batch`. Repeating this command with various values of h will allow you to examine visually the effect of choosing smaller and smaller step sizes.

We will also want quantitative information about the error. To get this kind of information we have to evaluate the exact solution at the same points at which we have the approximate values calculated, i.e. at the points in the vector `s`. This is easy. Simply enter `z=-1-s+2*exp(s);`. Then to compare with the approximate values, we look at the difference of the two vectors `w=z-y`. We are only interested in the magnitude of the error and not the sign. The command `abs(z-y)` gives us a vector with the absolute values of each entry of `x-y`. Finally, we only want to know the largest error, so the MATLAB command we need is `maxerror = max(abs(z-y))`. We enter that command without the semi-colon, because we want to see the result on the command window. Therefore, to effect these changes we alter the file `batch.m` to the following:

```
t=0:0.05:3;
x=-1-t+2*exp(t);               % The exact solution.
```

```
[s,y]=eul('ypt',0,3,1,h);        % The approximate solution.
plot(t,x,s,y,'o')                % Plot both.
z=-1-s+2*exp(s);                 % The exact solution at the s points.
maxerror = max(abs(z-y))         % The biggest error.
```

A typical session would result in the following on the command window:

```
>> h=0.2;batch

maxerror =

    9.3570

>> h=0.1;batch

maxerror =

    5.2723

>> h=0.05;batch

maxerror =

    2.8127

>> h=0.025;batch

maxerror =

    1.4548
```

Of course, each of these commands will result in a visual display of the error on the Figure Window as well. Thus, we can very easily examine the error in Euler's method both graphically and quantitatively.

If you have executed all of the commands up to now, you will have noticed that decreasing the step size has the effect of reducing the error. What is missing is a graphic display of how the error changes with step size. It would be illuminating to see a plot of the maximum error versus step size. To do this using MATLAB, we have to devise a way to capture the various step sizes and the associated errors into vectors, and then use one of the plotting commands.

We will call the vector of step sizes h_vect, and the vector of errors err_vect. It is necessary to initialize these vectors. Since at the beginning they contain no information, the proper command is h_vect=[];. The symbol [] stands for the empty matrix; err_vect is initialized in a similar manner.

Next it will be necessary to add new values of step size to h_vect as they are calculated. The command to do this is h_vect = [h_vect,h]. The result will be a row vector with one more entry than h_vect had previously. We will put all of these commands into an M-file. Let's call it batch2.m. The final contents will be:

```
t = 0:0.05:3;
x = -1-t+2*exp(t);              % The exact solution.
h_vect = [h_vect,h];            % The new step size vector.
[s,y] = eul('ypt',0,3,1,h);     % The approximate solution.
plot(t,x,s,y,'o')               % Plot both.
z = -1-s+2*exp(s);              % The exact solution at the s points.
maxerror = max(abs(z-y));       % The biggest error.
err_vect = [err_vect, maxerror];     % The new error vector.
h_vect                          % Display the vector of step sizes.
err_vect                        % Display the vector of errors.
h = h/2;                        % The step size for the next time.
```

Now we are ready to go. We enter the initialization step and then execute batch2.

```
>> h_vect=[];err_vect=[];h=0.2;
>> batch2

h_vect =

    0.2000

err_vect =

    9.3570
```

When we execute batch2 once more, we get longer vectors.

```
>> batch2

h_vect =

    0.2000    0.1000

err_vect =

    9.3570    5.2723
```

74

Continuing to execute `batch2`, we get more and more data. On the seventh iteration we get the following:

```
>> batch2

h_vect =

    0.2000    0.1000    0.0500    0.0250    0.0125    0.0063    0.0031

err_vect =

    9.3570    5.2723    2.8127    1.4548    0.7401    0.3733    0.1875
```

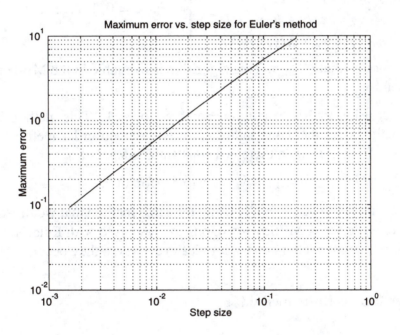

Figure 5.2. Error analysis for Euler's method.

Executing `batch2` one more time, so that we have eight data points in all, we are ready to plot the data. This can be done with any of the plotting commands. Try them all to see which you like best. We will use a loglog graph. Making the graph pretty requires a few more commands:

```
>> loglog(h_vect,err_vect)
>> xlabel('Step size')
>> ylabel('Maximum error')
```

```
>> title('Maximum error vs. step size for Euler''s method')
>> grid
```

Figure 5.2 shows the result.

The second order Runge-Kutta method

This method is sometimes called the improved Euler's method. To find this approximation to the solution of the initial value problem $y' = f(t, y)$, with $y(a) = y_0$, on the interval $[a, b]$, we set $t_0 = a$, we choose a step size h, and then we inductively define

$$\left. \begin{aligned} t_{k+1} &= t_k + h \\ s_1 &= f(t_k, y_k) \\ s_2 &= f(t_k + h, y_k + h s_1) \\ y_{k+1} &= y_k + h(s_1 + s_2)/2 \end{aligned} \right\} \quad \text{for } k = 0, 1, 2, \dots, N,$$

where N is large enough that $t_N \geq b$. This algorithm is available in the M-file rk2.m, which is listed in the Appendix.

The syntax for using rk2 is exactly the same as for eul. This method is more accurate than is Euler's method. As the name indicates, it is a second order method, i.e. if y_k is the calculated approximation at t_k, then there is a constant C such that

$$|y(t_k) - y_k| \leq C|h|^2 \quad \text{for all } k.$$

The error can be examined experimentally using the same method that we illustrated for Euler's method. In fact, to write script M-files analogous to those used with Euler's method, it is only necessary to replace the three letters eul with rk2 everywhere they occur.

The fourth order Runge-Kutta method

This is the final method we want to consider. To find this approximation to the solution to the initial value problem $y' = f(t, y)$, with $y(a) = y_0$, on the interval $[a, b]$, we set $t_0 = a$, we choose a step size h, and then we inductively define

$$\left. \begin{aligned} t_{k+1} &= t_k + h \\ s_1 &= f(t_k, y_k) \\ s_2 &= f(t_k + h/2, y_k + h s_1/2) \\ s_3 &= f(t_k + h/2, y_k + h s_2/2) \\ s_4 &= f(t_k + h, y_k + h s_3) \\ y_{k+1} &= y_k + h(s_1 + 2 s_2 + 2 s_3 + s_4)/6 \end{aligned} \right\} \quad \text{for } k = 0, 1, 2, \dots, N,$$

76

where N is large enough that $t_N \geq b$. This algorithm is available in the M-file rk4.m, which is listed in the Appendix.

The syntax for using rk4 is exactly the same as for eul or rk2. As the name indicates, it is a fourth order method, i.e. if y_k is the calculated approximation at t_k, then there is a constant C such that

$$|y(t_k) - y_k| \leq C|h|^4 \quad \text{for all } k.$$

Again, the error can be examined experimentally using the same method that we illustrated for Euler's method. This time replace eul (or rk2) with rk4 in the script M-files.

Comparing the three methods

It is very interesting to compare the accuracy of these three methods, both for individual step sizes, and for a range of step sizes. This can be done by writing script M-files that are only minor modifications of those already used in this chapter. Here is one that does the job. We are still working on the differential equation $x' = x + t$ with initial value $x(0) = 1$.

```
t = 0:0.05:3;
x = -1-t+2*exp(t);             % The exact solution.
h_vect = [h_vect,h];           % The new step size vector.
[s,y1] = eul('ypt',0,3,1,h);   % The approx. sol'n for Euler's method.
[s,y2] = rk2('ypt',0,3,1,h);   % The approx. solution for rk2.
[s,y4] = rk4('ypt',0,3,1,h);   % The approx. solution for rk4.
plot(t,x,s,y1,'o',s,y2,'+',s,y4,'x')          % Plot all.
z = -1-s+2*exp(s);             % The exact solution at the s points.
eulerror = max(abs(z-y1));     % The biggest error for eul.
rk2error = max(abs(z-y2));     % The biggest error for rk2.
rk4error = max(abs(z-y4));     % The biggest error for rk4.
eul_vect = [eul_vect, eulerror];     % The new error vector.
rk2_vect = [rk2_vect, rk2error];     % The new error vector.
rk4_vect = [rk4_vect, rk4error];     % The new error vector.
h_vect                         % Display the vector of step sizes.
eul_vect                       % Display the vectors of errors.
rk2_vect
rk4_vect
h = h/2;                       % The new step size.
```

This needs the initialization command

```
h=0.2;eul_vect=[];rk2_vect=[];rk4_vect=[];h_vect=[];
```

Finally the script can be executed several times in a row as before to accumulate data in the four vectors. Then the sequence of commands

```
>> loglog(h_vect,eul_vect,h_vect,rk2_vect,h_vect,rk4_vect)
>> grid
>> xlabel('Step size')
>> ylabel('Maximum error')
>> title('Maximum error vs. step size')
>> gtext('eul')
>> gtext('rk2')
>> gtext('rk4')
```

will result in a plot of the error curves for each of the three methods, as we see in Figure 5.3. Each of the `gtext` commands will put the indicated text at the point on the figure where the mouse button is clicked.

Notice that the curves are nearly straight lines in this loglog plot. The slopes are also roughly comparable to the order of the method in each case.

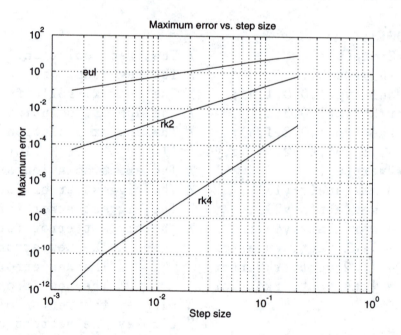

Figure 5.3. Comparison of errors for the three methods.

There is one anomaly that might have been noticed by the reader who has carried out the previous task. The vector `rk4_vect` has zero entries except for the first two. To see what the entries really are we need to print the vector with more accuracy. To do this, enter `format long` and then `rk4_vect` to see what you get.

78

Exercises

1. Consider the following initial value problems on the indicated intervals
    i) $x' = x \sin(3t)$, with $x(0) = 1$ on $[0, 4]$; $h_0 = 0.4$.
    ii) $y' = (1 + y^2) \cos(t)$, with $y(0) = 0$ on $[0, 6]$; $h_0 = 1$.
    iii) $z' = z^2 \cos(2t)$, with $z(0) = 1$ on $[0, 6]$; $h_0 = 0.5$.

    For each of these:

    a) Find the exact solution.

    b) Examine the error involved in solving this equation numerically using each of the three methods (i.e. `eul`, `rk2`, and `rk4`) with eight different step sizes starting with h_0, and halving each time. On a single sheet of graph paper plot the maximum error vs. step size curve for each of the three methods. A loglog graph would be preferable, but not necessary. Of course you can use MATLAB to do this job for you.

    c) Use MATLAB to plot on a single Graphics Window the graph of the exact solution, together with the plots of the solutions using each of the three methods with step size $h = h_0/4$. Use a distinctive marking for each method. Label the graph appropriately, and print the Graphics Window.

2. Notice that the graphs in Figure 5.3 are almost straight lines. What is the significance of the slopes of the three curves?

3. The accuracy of any numerical method in solving a differential equation $y' = f(t, y)$ depends on how strongly the equation depends on the variable y. More precisely, the constants that appear in the error bounds depend on the derivatives of f with respect to y. To see this experimentally, consider the two initial value problems

$$y' = y \qquad y(0) = 1,$$
$$y' = e^t \qquad y(0) = 1.$$

You will notice that the two problems have the same solution. For each of the three methods described in this chapter compute approximate solutions to these two initial value problems over the interval $[0, 1]$ using a step size of $h = 0.01$. For each method compare the accuracy of the solution to the two problems.

4. Remember that $y(t) = e^t$ is the solution to the initial value problem $y' = y$, $y(0) = 1$. Then $e = e^1$, and in MATLAB this is $e = \texttt{exp(1)}$. Suppose we try to calculate e approximately by solving the initial value problem, using the methods of this chapter. Use step sizes of the form $1/n$, where n is an integer. For each of Euler's method, the second order Runge-Kutte method, and the fourth order Runge-Kutte method, how large does n have to be to get an approximation e_{app} which satisfies $|e_{\text{app}} - e| \le 10^{-3}$?

5. In the previous problem, show that the approximation to e using Euler's method with step size $1/n$ is $(1 + 1/n)^n$. As a challenge problem, compute the formulas for the approximations using the two Runge-Kutte methods.

Appendix.   M-files for numerical methods

The first M-file is for Euler's method.

```
function [tout, yout] = eul(FunFcn, t0, tf, y0, ssize)

% EUL     Integrates a system of ordinary differential equations
%         using Euler's method.
%
%         [T,Y] = EUL('yprime', t0, tfinal, y0) integrates the
%         system of ordinary differential equations described
%         by the M-file YPRIME.M over the interval T0 to Tf and
%         using initial conditions Y0.
%         [T, Y] = EUL('yprime', t0, tf, y0, ssize) uses step
%         size ssize
%
% INPUT:
% yprime  Function M-file describing the equation.
% t0     - Initial value of t.
% tf     - Final value of t.
% y0     - Initial value vector.
% ssize - The step size. (Default: ssize = (tf - t0)/100).
%
% OUTPUT:
% T  - Returned integration time points (column-vector).
% Y  - Returned solution, one solution row-vector per t-value.

% Initialization

if (nargin < 5), ssize = (tfinal - t0)/100; end

h = ssize;
t = t0;
y = y0(:);
tout = t;
yout = y.';
% The main loop
while (t < tfinal)
        if t + h > tfinal, h = tfinal - t; end
        % Compute the slope
        s1 = feval(FunFcn, t, y); s1 = s1(:); % s1=f(t(k),y(k))
```

```
        t = t + h;
        y = y + h*s1;                  % y(k+1) = y(k) + h*f(t(k),y(k))
        tout = [tout; t];
        yout = [yout; y.'];
    end;
```

Since we are not teaching programming, we will not explain everything in this file, but a few things should be explained. The % is MATLAB's symbol for comments. MATLAB ignores everything that appears on a line after a %. The large section of comments that appears at the beginning of the file can be read using the MATLAB help command. Enter help eul and see what happens. Notice that these comments at the beginning take up more than half of the file. That's an indication of how easy it is to program these algorithms. In addition, a couple of lines of the code are needed only to allow the program to handle systems of equations.

Next, we present the M-file for the second order Runge-Kutta method.

```
function [tout, yout] = rk2(FunFcn, t0, tf, y0, ssize)

% RK2    Integrates a system of ordinary differential equations
%        using the second order Runge-Kutta  method.
%        [T,Y] = RK2('yprime', t0, tf, y0) integrates the system
%        of ordinary differential equations described by the
%        M-file YPRIME.M over the interval T0 to Tf and using
%        initial conditions Y0.
%        [T, Y] = RK2('yprime', t0, tf, y0, ssize) uses step
%        size ssize
%
% INPUT:
% yprime  Function M-file describing the equation.
% t0    - Initial value of t.
% tf    - Final value of t.
% y0    - Initial value column-vector.
% ssize - The step size. (Default: ssize = (tf - t0)/100).
%
% OUTPUT:
% T   - Returned integration time points (column-vector).
% Y   - Returned solution, one solution row-vector per t-value.


% Initialization

if nargin < 5, ssize = (tfinal - t0)/100; end
h=ssize;
```

```
t = t0; y = y0(:);
tout = t;
yout = y.';

% The main loop
while (t < tfinal)
      if t + h > tfinal, h = tfinal - t; end

      % Compute the slopes
      s1 = feval(FunFcn, t, y); s1 = s1(:);
      s2 = feval(FunFcn, t + h, y + h*s1); s2=s2(:);
      t = t + h;
      y = y + h*(s1 + s2)/2;
      tout = [tout; t];
      yout = [yout; y.'];

end;
```

Finally, we have the M-file for the fourth order Runge-Kutta method.

```
function [tout, yout] = rk4(FunFcn, t0, tf, y0, ssize)

% RK4    Integrates a system of ordinary differential equations
%        using the fourth order Runge-Kutta  method.
%        [T,Y] = RK4('yprime', t0, tf, y0) integrates the system
%        of ordinary differential equations described by the
%        M-file yprime.m over the interval t0 to tf and using
%        initial conditions y0.
%        [T, Y] = RK4('yprime', t0, tf, y0, ssize) uses step
%        size ssize
%
% INPUT:
% yprime  Function M-file describing the equation.
% t0     - Initial value of t.
% tfinal- Final value of t.
% y0     - Initial value column-vector.
% ssize - The step size. (Default: ssize = (tf - t0)/100).
%
% OUTPUT:
% T  - Returned integration time points (column-vector).
% Y  - Returned solution, one solution row-vector per t-value.
```

```
% Initialization

if nargin < 5, ssize = (tfinal - t0)/100; end
h=ssize;
t = t0; y = y0(:);
tout = t;
yout = y.';

% The main loop
while (t < tfinal)
        if t + h > tfinal, h = tfinal - t; end

% Compute the slopes
        s1 = feval(FunFcn, t, y); s1 = s1(:);
        s2 = feval(FunFcn, t + h/2, y + h*s1/2); s2=s2(:);
        s3 = feval(FunFcn, t + h/2, y + h*s2/2); s3=s3(:);
        s4 = feval(FunFcn, t + h, y + h*s3); s4=s4(:);
        t = t + h;
        y = y + h*(s1 + 2*s2 + 2*s3 +s4)/6;
        tout = [tout; t];
        yout = [yout; y.'];

end;
```

6.   Solving ODEs in MATLAB

This is one of the most important chapters in this manual. Here we will describe how to use MATLAB's own solvers to find approximate solutions to almost any system of differential equations. In addition to being very powerful, they are easy to use, as the reader will discover. The methods described herein are used regularly by engineers and scientists, and are available in any version of MATLAB. The student who learns the techniques described here will find them useful in many later circumstances.

As this Manual is going to print, a new suite of ode solvers is being developed for MATLAB. The first thing the reader should do is to find out whether the new suite is available or not. To do this, enter `help funfun` at the command line. MATLAB will print out a list of functions that act on other functions. Included in the list there will be some that solve differential equations. The list will certainly contain `ode23` and `ode45`. If, in addition, the list contains `ode113`, `ode15s`, or `ode23s`, then you have the new suite of solvers installed, otherwise you do not.*

In either case there will be a solver named `ode45`. This is the one we recommend for use with this Manual. Although the old and the new versions have the same name, the usage syntax is slightly different, and the two solvers operate slightly differently. We will carefully point out the difference in the syntax, but we will not explore the different internal workings of the routines.

Although we will almost exclusively use `ode45`, it is important to realize that the calling syntax is the same for the other solvers. I.e., in the old version `ode23` and `ode45` use the same syntax, and in the new version all five of the solvers use the same syntax. We will spend a paragraph discussing the use of `ode15s` at the end of the chapter.

Single first order differential equations

The basic syntax for using `ode45` is

```
[t,x] = ode45('yprime',t0,tf,y0);        % Old ode45
```

or

```
[t,x] = ode45('yprime',[t0,tf],y0);      % New ode45
```

---

* The MATLAB ODE suite is planned for release by The MathWorks in a later version of MATLAB. In the meantime the functions can be obtained by anonymous ftp from `ftp.math works.com`, in the directory `pub/mathworks/toolbox/matlab/funfun`.

Here we are solving the equation $y' = $ yprime(t, y), with the initial condition $y(t_0) = y_0$, on the interval $t_0 \leq t \leq t_f$. The description of yprime must be in a function M-file with the name yprime.m. The format for function M-files is described in Chapter 3; they need the same format for both versions of ode45 as they need for eul, rk2, or rk4.

Notice that the only difference in the basic syntax is that in the new version, the initial and final times t0 and tf must be put into a vector [t0,tf]. We will find other differences later, when we examine the options available with the two solvers.

Both versions of ode45 use variable step Runge-Kutta procedures. Six derivative evaluations are used to calculate an approximation of order five, and then another of order four. These are compared to come up with an estimate of the error being made at the current step. It is required that this estimated error at the kth step should be less than a predetermined amount. In the old version of ode45 this means that

$$|\text{estimated error}| \leq \max(1, |y_k|) \times \text{tolerance}, \tag{1}$$

where y_k is the calculated solution at step k. The default value of tolerance is 10^{-6}. In the new version of ode45 the estimated error is required to satisfy a similar, but more complicated inequality, which involves two tolerance parameters. We will discuss this later in this chapter. For the most of the uses in this Manual the default values of the tolerance parameters will provide sufficient accuracy.

The output of ode45 (both versions) consists of the column vector t, and the matrix x. The vector t is a list of the values of the variable t at which the approximate solution has been calculated. In the case of a single, first order equation, x is also a column vector. Each entry of x is the approximate solution to the differential equation at the corresponding value of t.

As an example, consider the equation $x' = \cos(t)/(2x - 2)$, with initial condition $x(0) = 3$, on the interval $[0, 2\pi]$. To solve this with ode45 we enter:

```
[t,x]=ode45('test',0,2*pi,3);          % Old version
```

or

```
[t,x]=ode45('test',[0,2*pi],3);          % New version
```

where test.m is the M-file which contains the definition of the equation, i.e.,

```
function xpr = test(t,x)

xpr = cos(t)/(2*(x-1));
```

To see what the answer looks like, we can enter t at the MATLAB prompt, and then enter x. We discover that both t and x are column vectors with the same number of entries. If you are using the new version of MATLAB, these vectors will probably be longer than can be displayed on a screen. In this case enter t(1:20) and x(1:20), which will display the first twenty entries of t and x.

A better way to see the relationship between t and x is to print them to the screen side by side. This can be accomplished by realizing that the command [t,x] will have MATLAB put the two column vectors into one matrix. Entering [t,x] (or [t(1:20), x(1:20)]) at the MATLAB prompt will print the two side by side. Try it.

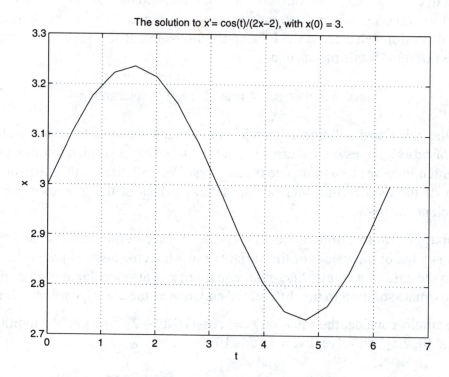

Figure 6.1. A solution using ode45.

Plotting the solution is quite easy. The following will do that and make the graph look pretty.

```
>> plot(t,x)
>> title('The solution to x''= cos(t)/(2x-2), with x(0) = 3.')
>> xlabel('t')
>> ylabel('x')
>> grid
```

The result (for the old version of ode45) is shown in Figure 6.1.

86

Systems of first order equations

Actually, systems are no harder to handle using `ode45` than are single equations. As an example, we will solve the system

$$x_1' = x_2 - x_1^2$$
$$x_2' = -x_1 - 2x_1x_2$$

numerically on the interval $[0, 10]$, with initial conditions $x_1(0) = 0$ and $x_2(0) = 1$. It is first necessary to describe the system in a derivative M-file. We will use the vector x in MATLAB to denote the solution. The first component, which in MATLAB is denoted by `x(1)`, will correspond to x_1. Similarly `x(2)` will correspond to x_2. For the derivatives we will use the vector `xpr`, with the first component corresponding to x_1', and the second to x_2'. Then our system of differential equations can be entered into an M-file as follows:

```
function xpr=test2(t,x);

xpr(1)=x(2)-x(1)^2;
xpr(2)=-x(1)-2*x(1)*x(2);
```

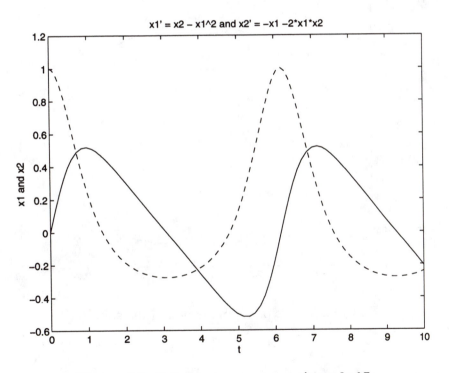

Figure 6.2. Solution to a system using `ode45`.

87

Now, to solve this system we use almost the same command as before:

```
[t,x]=ode45('test2',0,10,[0,1]);          % Old version.
```

or

```
[t,x]=ode45('test2',[0,10],[0,1]);        % New version.
```

The only difference is that for the initial value we must have a vector [0,1], which is the same as $(x_1(0), x_2(0))$.

The output vector t is similar to what it was in the earlier case. It is a vector containing the list of the t values where the approximate solutions were computed. Print it to your screen by entering t at the MATLAB prompt to see what it looks like. It will be quite long. You might want to check the size of t by entering size(t). To see the first twenty entries, enter t(1:20).

This time x is really a matrix instead of a column vector. To see the first twenty rows, enter x(1:20,:) at the MATLAB prompt. Each row of this matrix is a vector containing the approximate solution at the corresponding value of t. The first component of this row vector is the approximate value of x_1, and the second is that of x_2.

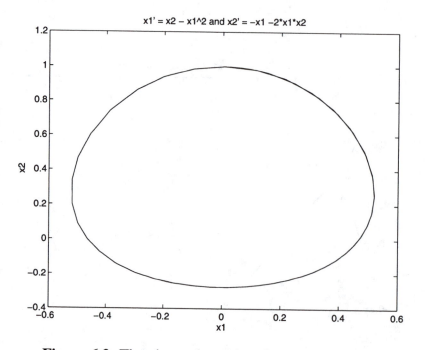

Figure 6.3. The phase plane plot of the solution.

To plot both components of the solution as functions of t, it is only necessary to enter plot(t,x). The results are shown in Figure 6.2. If only the first component of the solution

88

is wanted, enter `plot(t,x(:,1))`. The colon in the notation `x(:,1)` indicates that we want all rows, and the 1 that we want the first column. Similarly, if only the second component is wanted, enter `plot(t,x(:,2))`. This is an example of the very sophisticated subscripting options available in MATLAB.

It is also possible to plot the components against each other with the command

```
plot(x(:,1),x(:,2))
```

The result is called a *phase plane* plot, and it is shown in Figure 6.3.

Another way to present the solution to the system graphically is in a three dimensional plot, where both components of the solution are plotted as separate variables against t. MATLAB does this using the command `plot3`. For example, enter `plot3(t,x(:,1),x(:,2))` to see the plot with t along the x-axis, and the two components of the solution as y, and z. Alternatively, enter `plot3(x(:,1),x(:,2),t)` to see the solution with t along the z-axis, and the two components of the solution as x and y.

Second order differential equations

To solve a single second order differential equation it is necessary to replace it with the equivalent first order system. For the equation

$$y'' = f(t, y, y'), \tag{2}$$

we set $x_1 = y$, and $x_2 = y'$. Then $\mathbf{x} = (x_1, x_2)$ is a solution to the first order system

$$\begin{aligned} x_1' &= x_2 \\ x_2' &= f(t, x_1, x_2). \end{aligned} \tag{3}$$

More importantly, if $\mathbf{x} = (x_1, x_2)$ is a solution of the system in (3), we set $y = x_1$. Then we have $y' = x_1' = x_2$, and $y'' = x_2' = f(t, x_1, x_2) = f(t, y, y')$. Hence y is a solution of the equation in (2).

As a concrete example, consider the nonlinear equation $y'' + yy' + y = 0$, with initial conditions $y(0) = 0$, and $y'(0) = 1$. The corresponding system is

$$\begin{aligned} x_1' &= x_2 \\ x_2' &= -x_1 x_2 - x_1 \end{aligned}$$

Thus, we must create the M-file

```
function xpr=test3(t,x);

xpr(1)=x(2);
xpr(2)=-x(1)*x(2)-x(1);
```

and to solve the initial value problem on the interval [0, 10], we enter

```
>> [t,x]=ode45('test3',0,10,[0,1]);      % Old version.
```

or

```
>> [t,x]=ode45('test3',[0,10],[0,1]);    % New version.
```

To plot the solution y, we simply remember that $y = x_1$. Thus, $x(:,1)$ represents the solution, and we enter $plot(t,x(:,1))$.

The function ode45 is designed to solve systems of first order differential equations. We have just seen how a second order equation can be replaced by an equivalent first order system, and then solved. This method is completely general. Any system of ordinary differential equations can be replaced by an equivalent first order system. It is only necessary to add new variables for derivatives of the original variables, just as we did in this example. As a result, ode45 has the capability of solving almost any system of ordinary differential equations.

The Lorenz system

The solvers in MATLAB can solve first order systems containing as many equations as you like. As an example we will solve the Lorenz system. This is a system of three equations which was published in 1963 by the meteorologist and mathematician E. N. Lorenz. It represents a simplified model for atmospheric turbulence beneath a thunderhead. The equations are

$$\begin{aligned} x' &= -ax + ay \\ y' &= rx - y - xz, \\ z' &= -bz + xy \end{aligned} \tag{4}$$

where a, b, and r are positive constants.

In MATLAB we will use the vector u, where $u(1)$ corresponds to x, $u(2)$ to y, and $u(3)$ to z. The vector upr will be used for the corresponding derivatives. Then the derivative M-file lor.m could look like:

```
function upr = lor(t,u)

global AA  BB RR

upr(1) = -AA*u(1) + AA*u(2);
upr(2) = RR*u(1) - u(2) -u(1)*u(3);
upr(3) = -BB*u(3) + u(1)*u(2);
```

90

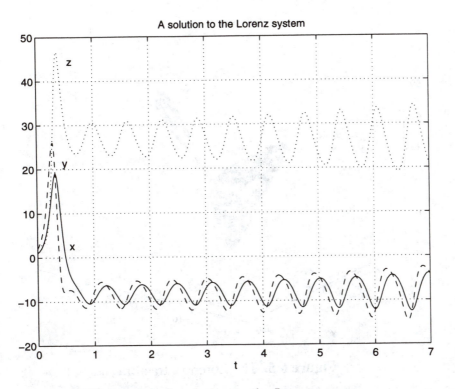

Figure 6.4. A solution to the Lorenz system.

(It would be natural to name this file `lorenz.m`, but there is already an M-file with that name. If you enter `lorenz` at the MATLAB prompt you will see a solution to the Lorenz system (4) displayed in a very attractive manner.)

The new feature here is the use of *global variables*. The symbols **AA**, **BB**, and **RR** correspond to the parameters a, b, and r in the system in (4). The line `global AA BB RR` declares them to be global variables, which means that, if we enter the same line in the Command Window, and then assign **AA**, **BB**, and **RR** values in the Command Window, these values will also be assigned to them whenever the derivative M-file `lor.m` is used. Hence we can change the values of the parameters in the Command Window, and we do not have to rewrite the M-file every time we want to make a change. For parameters in a differential system, this can save a lot of time.

We will start with $a = 10$, $b = 8/3$, and $r = 28$. Then to solve the Lorenz system over the interval [0, 7], with $x(0) = 1$, $y(0) = 2$, and $z(0) = 3$, we proceed as follows:

```
>> global AA  BB  RR
>> AA = 10;BB=8/3;RR=28;
>> [t,u] = ode45('lor',[0,7],[1,2,3]);
```

We plot the solution with the command `plot(t,u)`, and after some labeling we get Figure 6.4.

91

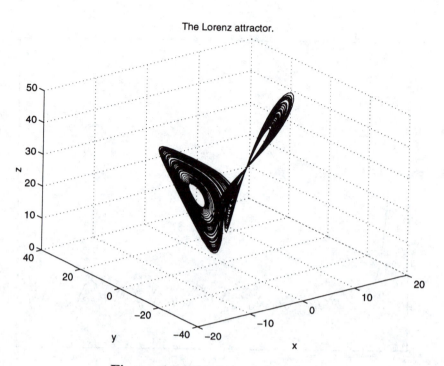

The Lorenz attractor.

Figure 6.5. The Lorenz attractor.

In Figure 6.4, there appears to be transient behavior in the interval $0 \le t \le 1$. Let's compute the solution over a longer period, say $0 \le t \le 20$. (If your computer is very fast, use the upper bound of 100 instead of 20.) We will then plot the part corresponding to $t > 1$ in three dimensions. This is easily accomplished. The required commands are:

```
>> [t,u] = ode45('lor',[0,100],[1,2,3]);      % New version.
>> N= find(t>1);
>> v=u(N,:);
>> plot3(v(:,1),v(:,2),v(:,3))
```

Some explanation is necessary. First, the command `N= find(t>1);` produces a list of the indices of those elements of `t` which satisfy $t > 1$. Then `v=u(N,:);` produces a matrix containing the rows of `u` with indices in `N`. The result is that `v` consists of only those solution points corresponding to $t > 1$, and the resulting plot represents the long time behaviour of the solution after the transient effects are gone.

After a little grooming we get Figure 6.5, a fine picture of the Lorenz attractor. The Lorenz system, with this particular choice of parameters, has the property that any solution, no matter what its initial point, is attracted to this rather complicated, butterfly-shaped, set. We will return to this in the exercises. There we will also examine the Lohrenz system for different values of the parameters.

Computational problems and some solutions

Both versions of the routine `ode45` are extremely flexible, fast, and accurate. They are good general purpose solvers. However, the computational solution of all differential equations cannot be achieved with just one solution method. Here we will discuss some of the problems that can arise. We will also indicate how to solve some of these problems.

How accurate is `ode45` really?

This is a very good question. The earlier description of how the old version of the routine works (see (1)) could lead one to believe that an estimate of the sort

$$|\text{error}| \leq \max(1, |\mathbf{y}|) \times \text{tolerance}$$

is valid. It would be wonderful if this were true, and for many examples it is true. Still there are examples that show different results, even over intervals of relatively short length. It is a good idea to introduce a safety factor, so a more reliable estimate of the error would be something like

$$|\text{error}| \leq 100 \times \max(1, |\mathbf{y}|) \times \text{tolerance}. \tag{5}$$

The user can use this estimate to decide which tolerance should be used to achieve a desired upper bound on error.

To change the tolerance in the old version of `ode45`, it is only necessary to add it as an extra parameter in the calling syntax. For example,

```
[t,x] = ode45('test',0,30*pi,3,1e-8);
```

would set the tolerance to 10^{-8}.

The new version of `ode45` takes a more sophisticated approach to error checking. If $\mathbf{y}^k$ is the computed solution at step k, then each component of the solution is required to satisfy its own error restriction. This means that we consider an error vector, which has a component for every component of $\mathbf{y}$, and it is required that each component of the error vector satisfy one of the following two inequalities:

$$|\text{estimated error}_j^k| \leq \max(|y_j^k|, |y_j^{k-1}|) \times \text{rtol}$$

$$|\text{estimated error}_j^k| \leq \text{atol}_j$$

Here rtol is called the *relative tolerance*, and **atol** is the *absolute tolerance*. Notice that the absolute tolerance is a vector quantity, with a component for each equation in the system being solved.

Just as before, the user can use these inequalities to determine the values of rtol and **atol** which are required to achieve the desired accuracy. The default values are rtol $= 10^{-3}$, and each component of **atol** $= 10^{-6}$.

To change the tolerances, it is necessary to use an options vector, which we will denote by `opt`. There are a large number of possible options for `ode45` and the other solvers in the new suite. To change any of them from the default values, we use the command `odeset`. For example, in the solution to the Lorenz system, we might want to use `rtol = 1e-4`, and `atol = [ 1e-5, 1e-6, 1e-6]`. To accomplish this we enter

```
opt = odeset('rtol',1e-4,'atol',[1e-5, 1e-6, 1e-6]);
[t,u] = ode45('lor',[0,20],[1,2,3],opt);
```

If we only wanted to change rtol, the first line should be

```
opt = odeset('rtol',1e-4);
```

In other words, we only specify those options which need to be changed from their default values.

The user has to be especially careful in using the estimate (5) for the old version of `ode45` with systems of equations. Different components of the solution vector might have markedly different sizes, and one or more of the components might be lost in the estimate. The term $|\mathbf{y}|$ will be dominated by the largest components of the vector $\mathbf{y}$, and the term $|\mathbf{error}|$ will likewise be dominated by the largest components of the vector **error**. The error made in calculating the smallest component might be very large relative to the size of the smallest component, and still be small in comparison to the errors in the larger components. This possibility is removed by the more sophisticated error estimation in the new version of `ode45`.

There is still a way in which excessive error can sneak up on the user. Consider the situation where it is the difference of two components of the solution which is really important. In that case each component has to be calculated to greater accuracy to ensure that the difference is sufficiently accurate.

The considerations to this point were for solutions over "relatively short" intervals of the independent variable. When solutions are required over long intervals, we have a whole new ball game. No solver can give reliable results over long intervals for all equations. For example, many equations have regions where solutions are extremely sensitive to initial conditions. This means that two solutions with initial conditions which are very close can eventually move very far apart. From a computational point of view, it must be recognized that small errors are being made all the time. Even if the error is very small, in a region of extreme sensitivity to initial conditions, the error might be enough to move to a solution which is diverging from the one sought. This is in the nature of differential equations, and cannot be anticipated by a solver without the intervention of the user.

When computing over a relatively large interval, no solution should be accepted uncritically. Compare the solution to what is expected. If nothing else, compute the solution again with a decreased tolerance, and compare the two solutions. For the new version of `ode45`, there is another option that is sometimes useful when computing over large intervals. This is `hmax`, the maximum allowed integration step. The default value is `(tf-t0)/10` where `t0` is the initial time, and `tf` is the final time. Reducing the value of `hmax` can sometimes assist the computation.

Behavior near singularities

Any numerical method will run into difficulties where the equation and/or the solution to the equation has a singularity, and `ode45` is no exception. There are at least three possible outcomes when `ode45` meets a singularity.

1. `ode45` can integrate right through the singularity, not even realizing it is there. In this case the accuracy of the result is highly doubtful, especially in that range beyond the singularity. This phenomenon will happen, for example, with the initial value problem $x' = x/(t - 1)$, with initial value $x(0) = 1$ on the interval $[0, 2]$.

2. The old version of `ode45` can find the singularity and report it with a comment like

   ```
   Singularity likely at t = 0.972656
   ```

   What this comment really means is that an extremely small step size was required to achieve the error limit. This is usually a sign of a singularity but not always. The new version is more circumspect. Its comment is

   ```
   Step failure at 1.129671e+01 with a minimum step size of 1.003350e-13
   ```

   Either of these messages probably indicate the presence of a singularity in the solution. An example of this is $x' = x/(1 - t)$, with initial value $x(0) = 1$ on the interval $[0, 1]$. One very nice thing about `ode45` is that if this happens, the output up to the point where the routine stops is made available. For example, if you execute `[t,w] = ode45('gizmo',[0,3.4],[0 1 2]);`, and the computation stops because a step size is called for which is smaller than the minimum allowed, then the variables `t` and `w` will contain the results of the computation up to that point.

3. `ode45` can choose smaller and smaller step sizes upon the approach to a singularity and go on calculating for a very long time — hours in some cases. For cases like this, it is important for the user to know how to stop a calculation in MATLAB. On most computers the combination of the control key and C depressed simultaneously will do the trick.

Kinky plots

This is a problem with the old verison of `ode45`. It can sometimes decide to use large step sizes, perhaps as large as the maximum allowed, which is hmax $= (t_f - t_0)/16$. Conceivably a computation could be completed in as few as 16 steps. Together with the initial point, this gives us only 17 data points. A curve plotted in MATLAB using 17 points usually looks smooth, but sometimes the result is a little kinky. As an example, the graph in Figure 6.1 is not as smooth as might be desired. The easiest solution for this problem is to use a smaller tolerance, which will force the use of a smaller step size, and will provide more data points.

The new version of `ode45` almost never has a problem with kinky plots. It can take as few as 10 steps to compute a solution, but the routine interpolates between the solution at the

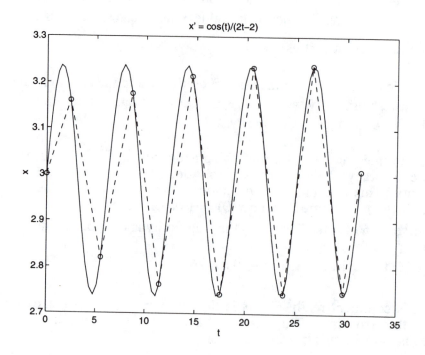

Figure 6.6. The effect of the refine option.

step points. The interpolation process uses the differential equation to compute the interpolating points, so they remain almost as accurate as the solution at the step points.

The number of interpolating points is controlled by the option `refine`. The default value of `refine` is 4. This should be sufficient to insure that most solution curves are smooth. If it should happen that a solution curve is not smooth, the solution can be recomputed with a larger value of `refine`. As an example consider the first equation in this chapter, $x' = \cos(t)/(2t - 2)$. This time we solve it over the interval $[0, 10\pi]$. The solution is periodic with period 2π, so we get five periods. One would expect the graph of such a function to be kinky, unless a large number of points were plotted. Indeed, the default value of `refine` does lead to a somewhat kinky graph of the solution. So, let's set `refine = 8`. The following list of commands

```
>> [t,x] = ode45('test',[0,10*pi],3,odeset('refine',8));
>> l=(length(t)-1)/8;          % The number of steps.
>> tp=t(8*[0:l]+1);            % The values of t at the step points.
>> xp=x(8*[0:l]+1);            % The values of x at the step points.
>> plot(t,x,tp,xp,'o',tp,xp,'--')
```

results in Figure 6.6. In this figure, the solid curve is the plot of the entire solution, and the circles represent the solution at the step points. Consequently, the dashed curve is what we would get without the refine option.

96

Stiff equations

Solutions to differential equations often have components which are varying at different rates. If these rates differ by a couple of orders of magnitude, the equations are called stiff. Stiff equations will not arise in this manual, but we will say a word about them in case the reader should have to deal with one.

An example of a stiff system is the van der Pol equation

$$x' = y$$
$$y' = -x + \mu(1 - x^2)y$$

when the parameter μ is very large, say $\mu = 1000$.

Runge-Kutta solution methods do not work well with stiff equations. The fast varying component requires a Runge-Kutta algorithm to use very small steps — so small that it takes a prohibitively long time to compute the solution. The new suite of ode solvers contains two routines, ode15s and ode23s which are designed to solve stiff equations. ode15s is the first of these to try.

A very attractive feature of the new suite of solvers is that the calling syntax is the same for all of them. Thus the syntax for using ode15s is the same as we have been using with ode45. For example,

```
[t,x] = ode15s('vdpol',tspan,init);
```

will compute the solution of the van der Pol system, if the system is described in the function M-file vdpol.m, tspan = [t0,tf] is the interval of computation, and init = [x0,y0] is vector of initial conditions. These routines also take an options vector as an additional parameter, and the makeup of that is again the same as it is for ode45. In particular the same options are available. There is one change — the default value of refine is 1.

But how do we tell if a system is stiff? It is often obvious from the physical situation being modeled that there are components of the solution which vary at rates that are significantly different, and therefore the system is stiff. Sometimes the presence of a large parameter in the equations is a tip off that the system is stiff. However, there is no general rule that allows us to recognize stiff systems. A good operational rule is to try to solve using ode45. If that fails, or is very slow, try ode15s.

What about solving in the negative direction in t?

This is simply not allowed by the old version of ode45. Furthermore, there does not seem to be an easy way around it. If it is necessary to solve in the negative direction, rewrite the differential equation by replacing t with $-t$, and then solve as usual. Of course it will then be necessary to change the sign of the vector t.

On the other hand, the new version can solve in the negative direction without a hitch.

Exercises

1. Use `ode45` to solve the following initial value problems numerically over the indicated intervals. Superimpose the solutions with different initial values on the same time plot.

    a) $x' = x^2 - t$, $x(0) = 0, 0.5, -0.5$, $0 \le t \le 4$.

    b) $x' = \cos(x)/(2 + \cos(t))$, $x(0) = 0, 1, \pi/2, 4, -\pi/2, -2$, $0 \le t \le 15$.

2. Use `ode45` to compute a solution to the initial value problem $x' = x/(1 + x)$, $x(0) = 1$ on the interval $[0, 3]$. Plot the solution. This equation cannot be solved explicitly, but since it is separable an implicit solution can be found. Check how closely the computed solution satisfies the implicit equation.

3. Using `ode45` calculate the solutions to the differential equation $yy'' - (y')^2 - y^2 = 0$ with each of the following initial conditions. In each case make a time plot and a phase plane plot of the results.

    a) $y(0) = e$, $y'(0) = -e$.

    b) $y(0) = 1$, $y'(0) = -1$.

    c) $y(0) = 1$, $y'(0) = 0$.

    d) $y(0) = 1$, $y'(0) = 3$.

    e) Find the general solution to the equation $yy'' - (y')^2 - y^2 = 0$. (Hint: Calculate $(y'/y)'$.)

4. The equation $x'' = ax' - b(x')^3 - kx$ was devised by Lord Rayleigh to model the motion of the reed in a clarinet. With $a = 5$, $b = 4$, and $k = 5$, solve this equation numerically with initial conditions $x(0) = A$, and $x'(0) = 0$ over the interval $[0, 10]$ for the three choices $A = 0.5$, 1, and 2. Prepare both time plots and phase plane plots containing the solutions to all three initial value problems superimposed. Describe the relationship that you see between the three solutions.

5. Consider the system of equations

$$x_1' = x_2$$
$$x_2' = -x_1$$

with initial conditions $x_1(0) = 1$, $x_2(0) = 0$. Find the exact solution. Then solve the system numerically over the interval $[0, 4\pi]$ using both `eul` with step size $h = 0.01$ and `ode45` with the default tolerance. Plot the phase plane for both solutions. This will give you a different picture of the errors made in the two methods.

6. **A predator-prey population model.**

In the 1920's, the Italian mathematician Umberto Volterra proposed the following mathematical model of a predator-prey situation to explain why, during the first World War, a larger percentage of the catch of Italian fishermen consisted of sharks and other fish eating fish than was true both before and after the war. Let $x(t)$ denote the population of the prey, and let $y(t)$ denote the population of the predators.

In the absence of the predators, the prey population would have a birth rate greater than its death rate, and consequently would grow according to the exponential model of population growth, i.e. the growth rate of the population would be proportional to the population itself. The presence of the predator population has the effect of reducing the growth rate, and this reduction depends on the number of encounters between individuals of the two species. Since it is reasonable to assume that the number of such encounters is proportional to the number of individuals of each population, the reduction in the growth rate is also proportional to the product of the two populations, i.e., there are constants a and b such that

$$x' = ax - bxy. \tag{6}$$

Since the predator population depends on the prey population for its food supply it is natural to assume that in the absence of the prey population, the predator population would actually decrease, i.e. the growth rate would be negative. Furthermore the (negative) growth rate is proportional to the population. The presence of the prey population would provide a source of food, so it would increase the growth rate of the predator species. By the same reasoning used for the prey species, this increase would be proportional to the product of the two populations. Thus, there are constants c and d such that

$$y' = -cy + dxy. \tag{7}$$

a) A typical example would be with the constants given by $a = 0.4$, $b = 0.01$, $c = 0.3$, and $d = 0.005$. Start with initial conditions $x_1(0) = 50$ and $x_2(0) = 30$, and compute the solution to (3) and (4) over the interval $[0, 100]$. Prepare both a time plot and a phase plane plot.

After Volterra had obtained his model of the predator-prey populations, he improved it to include the effect of "fishing", or more generally of a removal of individuals of the two populations which does not discriminate between the two species. The effect would be a reduction in the growth rate for each of the populations by an amount which is proportional to the individual populations. Furthermore, if the removal is truly indiscriminate, the proportionality constant will be the same in each case. Thus the model in equations (3) and (4) must be changed to

$$\begin{aligned} x' &= ax - bxy - ex \\ y' &= -cy + dxy - ey \end{aligned} \tag{8}$$

where e is another constant.

b) To see the effect of indiscriminate reduction, compute the solutions to the system in (5) when $e = 0, 0.01, 0.02, 0.03$, and 0.04, and the other constants are the same as they were in part a). Plot the five solutions on the same phase plane, and label them properly.

c) Can you use the plot you constructed in part b) to explain why the fishermen caught more sharks during World War I? You can assume that because of the war they did less fishing.

7. **Harmonic motion.**

The equation for the motion of a spring is

$$my'' + cy' + ky = F(t),$$

where m is the mass, c is the damping constant, and k is the spring constant. $F(t)$ represents the external force. For the following exercises, assume that $m = 1$kg, and that the spring constant $k = 16$N/m. We will be concerned with unforced oscillations, so we are are assuming that $F(t) = 0$. In each of the following cases compute the solution with initial conditions $x(0) = 1$, and $x'(0) = 0$ over the interval $[0, 20]$. Prepare both a time plot and a phase plane plot.

a) (No damping) $c = 0$.

b) (Under damping) $c = 2$.

c) (Critical damping) $c = 8$.

d) (Over damping) $c = 10$.

8. **The non-linear spring and Duffing's equation.**

A more accurate description of the motion of a spring is given by *Duffing's equation*

$$my'' + cy' + ky + ly^3 = F(t).$$

Here m is the mass , c is the damping constant, k is the spring constant, and l is an additional constant which reflects the "strength" of the spring. Hard springs satisfy $l > 0$, and soft springs satisfy $l < 0$. As usual, $F(t)$ represents the external force.

Duffing's equation cannot be solved analytically, but we can obtain approximate solutions numerically in order to examine the effect of the additional term ly^3 on the solutions to the equation. For the following exercises, assume that $m = 1$ kg, that the spring constant $k = 16$ N/m, and that the damping constant is $c = 1$ kg/sec. The external force is assumed to be of the form $F(t) = A\cos(\omega t)$, measured in Newtons, where ω is the frequency of the driving force. The natural frequency of the spring is $\omega_0 = \sqrt{k/m} = 4$ rad/sec.

a) Let $l = 0$ and $A = 10$. Compute the solution with initial conditions $y(0) = 1$, and $y'(0) = 0$ on the interval [0, 20], with $\omega = 3.5$ rad/sec. Print out a graph of this solution. Notice that the steady state part of the solution dominates the transient part when t is large.

b) With $l = 0$ and $A = 10$, compute the amplitude of the steady state solution as follows. The amplitude is the maximum of the values of $|y(t)|$. Because we want the amplitude of the steady state oscillation, we only want to allow large values of t, say $t > 15$. This will allow the transient part of the solution to decay. To compute this number in MATLAB, do the following. Suppose that Y is the vector of y-values, and T is the vector of corresponding values of t. At the MATLAB prompt type

```
max(abs(Y.*(T>15)))
```

Your answer will be a good approximation to the amplitude of the steady state solution.

Why is this true? The expression (T>15) yields a vector the same size as T, and an element of this vector will be 1 if the corresponding element of T is larger than 15 and 0 otherwise. Thus Y.*(T>15) is a vector the same size as T or Y with all of the values corresponding to $t \le 15$ set to 0, and the other values of Y left unchanged. Hence, the maximum of the absolute values of this vector is just what we want.

Set up a script M-file that will allow you to do the above process repeatedly. For example, if the derivative M-file for Duffing's equation is called duff.m, the script M-file could be

```
[t,y]=ode45('duff',0,20,[1,0]);
y = y(:,1);
amplitude = max(abs(y.*(t>15)))
```

Now by changing the value of ω in duff.m you can quickly compute how the amplitude changes with ω. Do this for eight evenly spaced values of ω between 3 and 5. Use plot to make a graph of amplitude vs. frequency. For approximately what value of ω is the amplitude the largest?

c) Set $l = 1$ (the case of a hard spring) and $A = 10$ in Duffing's equation and repeat b). Find the value of ω, accurate to 0.2 rad/sec, for which the amplitude reaches its maximum.

d) For the hard spring in c), set $A = 40$. You are to redo c), but with two different choices of initial conditions, and for eight evenly spaced values between 5 and 7. The initial conditions are $y(0) = y'(0) = 0$, and $y(0) = 6$, $y'(0) = 0$. Plot the two graphs of amplitude vs. frequency on the same figure. (The phenomenon you will observe is called Duffing's hysteresis.)

e) Set $l = -1$ and $A = 10$ (the case of a soft spring) in Duffing's equation and repeat c).

9. **The Lorenz system.**

   The purpose of this exercise is to explore the complexities displayed by the Lorenz system as the parameters are varied. We will keep $a = 10$, and $b = 8/3$, and vary r. For each value of r, examine the behavior of the solution with different initial conditions, and make conjectures about the limiting behavior of the solutions as $t \to \infty$. The solutions should be plotted in a variety of ways in order to get the information. These include time plots, such as Figure 6.4, and phase space plots, such as Figure 6.5. You might use plots of z versus x, etc. You might also use the techniques of the following exercise.

   Examine the Lorenz system for a couple of values of r in each of the following intervals. Describe the limiting behavior of the solutions.

   a) $0 < r < 1$.

   b) $1 < r < 470/19$. There are two distinct cases.

   c) $470/19 < r < 130$. The case done in the text. This is a region of chaotic behavior of the solutions.

   d) $150 < r < 165$. Things settle down somewhat.

   e) $215 < r < 280$. Things settle down even more.

10. Viewing the Lorenz attractor in three dimensions, as we do in Figure 6.5, is less than completely satisfactory. We really need to view the set from several different viewpoints in order to see what it really looks like. This is easily possible in MATLAB. The command `view` allows us to change the viewpoint of a three dimensional plot. The syntax is `view(azimuth, elevation)`, where elevation is the angle of the viewpoint above the (x, y) plane, and azimuth is the angle of the viewpoint in the (x, y) plane, measured counterclockwise from the x-axis. Both angles are measured in degrees. Entering `view(az,el)` causes the plot to be redone from the new viewpoint.

    a) Look at a Lorenz attractor from a variety of viewpoints, and print the two you think are most revealing.

    This process can be carried further using MATLAB's movie making capability. We will describe how to have MATLAB display a movie of the attractor rotating about the x-axis. The key is to create the following function M-file, and save it as `rotmovie.m`.

```
function M = rotmovie(u)

M = moviein(18);
v = sqrt(u(:,1).^2+u(:,2).^2);
mm = max(v);
lim=[-mm,mm,-mm,mm,min(u(:,3)),max(u(:,3))];
black=[0 0 0];
for j=1:18
        th = -pi+j*pi/9;
        A = [cos(th),sin(th),0;-sin(th),cos(th),0;0,0,1];
        uu=u*A;
        plot3(uu(:,1),uu(:,2),uu(:,3));
        axis(lim);
        set(gca,'xcolor',black,'ycolor',black,'zcolor',black);
        M(:,j)=getframe;
end
```

   If **u** is a matrix containing the x, y, and z components of the solution to the Lorenz system, then the command `M=rotmovie(u)` will create the data needed for the movie in the matrix M. Then the command `movie(M,10,6)` will repeat the movie 10 times at the rate of 6 frames per second.

b) Make a movie of the Lorenz attractor.

Notice that `rotmovie` can be used with any three dimensional data. It's use is not limited to the Lorenz attractor.

The M-file `rotmovie.m` needs some explanation, but we choose not to provide it. Instead we invite the reader to do some reverse engineering with the help of the MATLAB `help` command, the MATLAB *User's Guide,* and the MATLAB *Reference Guide.*

11. **Motion near the Lagrange points.**

Consider two large spherical masses of mass $M_1 > M_2$. In the absence of other forces, these bodies will move in elliptical orbits about their common center of mass (if it helps, think of these as the earth and the moon). We will consider the motion of a third body (perhaps a spacecraft), with mass which is negligible in comparison to M_1 and M_2, under the gravitational influence of the two larger bodies. It turns out that there are five equilibrium points for the motion of the small body relative to the two larger bodies. Three of these were found by Euler, and are on the line connecting the two large bodies. The other two were found by Lagrange and are called the Lagrange points. Each of these forms an equilateral triangle in the plane of motion with the positions of the two large bodies. We are interested in the motion of the spacecraft when it starts near a Lagrange point.

In order to simplify the analysis, we will make some assumptions, and choose our units carefully. First we will assume that the two large bodies move in circles, and therefore maintain a constant distance from each other. We will take the origin of our coordinate system at the center of mass, and we will choose rotating coordinates, so that the x-axis always contains the two large bodies. Next we choose the distance between the large bodies to be the unit of distance, and the sum of the the the two masses to be the unit of mass. Thus $M_1 + M_2 = 1$. Finally we choose the unit of time so that a complete orbit takes 2π units; i.e., in our units, a year is 2π units. This last choice is equivalent to taking the gravitational constant equal to 1.

With all of these choices, the fundamental parameter is the relative mass of the smaller of the two bodies

$$\mu = \frac{M_2}{M_1 + M_2} = M_2.$$

Then the location of M_1 is $(-\mu, 0)$, and the position of M_2 is $(1 - \mu, 0)$. The position of the Lagrange point is $((1 - 2\mu)/2, \sqrt{3}/2)$. If (x, y) is the position of the spacecraft, then the distances to M_1 and M_2 are

$$r_1^2 = (x + \mu)^2 + y^2,$$
$$r_2^2 = (x - 1 + \mu)^2 + y^2.$$

Finally, Newton's equations of motion in this moving frame are

$$
\begin{aligned}
x'' - 2y' - x &= -(1 - \mu)(x + \mu)/r_1^3 - \mu(x - 1 + \mu)/r_2^3, \\
y'' + 2x' - y &= -(1 - \mu)y/r_1^3 - \mu y/r_2^3.
\end{aligned}
\tag{9}
$$

a) Find a system of four first order equations which is equivalent to (9).

b) If the two bodies are the earth and the moon, $\mu = 0.0122$. The Lagrange points are stable for $0 < \mu < 1/2 - \sqrt{69}/18 \approx 0.03852$, and in particular for the earth/moon system. Write $x = (1 - 2\mu)/2 + \xi$, and $y = \sqrt{3}/2 + \eta$, so (ξ, η) is the position of the spacecraft relative to the Lagrange point. Starting with initial conditions which are less than 1/2 unit away from the Lagrange point, compute the solution. For each solution that you compute, make a plot of η vs. ξ to see the motion relative to the Lagrange point, and a plot of y vs. x, which also includes the positions of M_1 and M_2 to get a more global view of the motion.

c) Examine the range of stability by computing and plotting orbits for $\mu = 0.037$ and $\mu = 0.04$.

d) What is your opinion? Assuming μ is in the stable range, are the Lagrange points just stable, or are they asymptotically stable?

e) Find all five equilibrium points for the system you found in a). This is an algebraic problem of medium difficulty. It is not a computer problem unless you can figure out how to get the Symbolic Toolbox to find the answer.

f) Show that the equilibrium points on the x-axis are always unstable. This is a hard algebraic problem.

g) Show that the Lagrange points are stable for $0 < \mu < 1/2 - \sqrt{69}/18$. Decide whether or not these points are asymtotically stable. This is a very difficult algebraic problem.

7. More About the Symbolic Toolbox

In this chapter we revisit the Symbolic Toolbox. We will talk further about the commands `solve` and `dsolve`, and we will give some examples of how the Symbolic Toolbox can be used to do algebraic manipulations.

We have already seen that the Symbolic Toolbox and other symbolic algebra programs provide an easy to use tool for solving simple equations, either of the algebraic or differential variety. They can also be extremely useful in doing rather complex algebra, but this latter usage does not come without some effort on the part of the user. In order to take full advantage of the capabilities of the Symbolic Toolbox, it is necessary that the user be very good at algebraic manipulations, and very familiar with the capabilities of the program.

In this chapter, we will not always show the results of our calculations. It will be helpful if you enter commands without the semicolon, and, if necessary, use the command `pretty` to display the results. If a particular expression is still too complicated, you might try to use `simple`.

The answers you get in response to some of the commands may be slightly different from what is printed here. This will consist of differing orders of the terms in an algebraic expression. If this happens, check to be sure that your answer represents the same algebraic expression as is printed here. These kinds of differences are unavoidable, since different versions of MATLAB give slightly different responses to the commands in the Symbolic Toolbox.

Higher order differential equations

MATLAB and `dsolve` can be used to solve higher order differential equations, and even systems of equations. However, it is easier to stump MATLAB in these cases. The usage of `dsolve` is the same as before, but we do need to know that in MATLAB, the second derivative of y is denoted by D2y. Similarly the third derivative is D3y, etc.

Let's try to solve the differential equation $yy'' - (y')^2 - y^2 = 0$.

```
>> y = dsolve('y*D2y-(Dy)^2-y^2=0')

y =

exp(-1/2*C1+1/2*x^2+x*C2+1/2*C2^2)
```

First of all, notice that since there was no independent variable in the equation, MATLAB uses the default variable x. Next, notice that there are two constants $C1$ and $C2$ present, as we would expect, since we are solving a second order equation.

Finally, notice that MATLAB seems to have made a subtle mistake that many people make. The solution given here is an exponential, and, therefore, presumably always takes on positive values. On the other hand, if $z(t) = -y(t)$, then it is easy to check that z is also a solution to the differential equation. The fact is that MATLAB allows the constants to be complex numbers, and the exponential of a complex number is not usually positive.

To see how this works, let's solve the initial value problem for this equation with initial conditions $y(0) = -1$, and $y'(0) = 1$. The negative initial value for y means that MATLAB will have to adjust to the problem pointed out in the previous paragraph.

We need to know how to enter the second initial condition. In MATLAB's language this translates to Dy(0)=1, and it becomes a third equation to be entered into dsolve.

```
>> y = dsolve('y*D2y-(Dy)^2-y^2=0','y(0)=-1','Dy(0)=1')

y =

exp(i*pi+1/2*x^2-x)
```

We see that MATLAB is using a complex number in the exponent. By Euler's formula, we know that
$$e^{\pi i} = \cos(\pi) + i\sin(\pi) = -1,$$
so the solution is
$$y(x) = -e^{x^2/2-x}.$$

Is MATLAB smart enough to know this? We will get some good information if we use the command simple without assigning the result to a variable.

```
>> simple(y)

expand:

-exp(x^2)^(1/2)/exp(x)

ans =

exp(i*pi+1/2*x^2-x)
```

We see that MATLAB tried the command expand, and in response got the expression -exp(x^2)^(1/2)/exp(x). Although this is clearly equal to the expression we found, MATLAB discarded it because the length of this expression is longer than the original one. If we wanted to have y in this form, we could use the hint provided by simple and execute y = expand(y).

The forced undamped vibrating spring

In order to see how MATLAB can be used to derive the solutions of differential equations and their properties, let's examine the vibrating spring under the action of a periodic force. The equation of motion is

$$my'' + dy' + ky = A\cos(\omega t). \tag{1}$$

In this equation y is the distance of the object at the end of the spring from its equilibrium position, m is the mass of the body, k is the spring constant which describes the restoring force of the spring, and d is the damping constant. It is assumed that there is an external force of the form $A\cos(\omega t)$, where A is the amplitude and ω is the frequency.

It is important to realize that equation (1) also arises in the study of simple electrical circuits containing only a capacitor, a resistor, an inductance and a voltage source in series. The only difference is that the symbols used in equation (1) are not very appealing for that application, since y represents the charge across a capacitor, m is the inductance, d is the resistance, and k is the reciprocal of the capacitance.

We will look first at the case where there is no damping, i.e. $d = 0$. We will also assume that at time $t = 0$ the object is at rest, so $y(0) = 0$ and $y'(0) = 0$. In this case MATLAB can solve the equation. (We must remember that the independent variable is t.)

```
>> y=dsolve('m*D2y+k*y=A*cos(w*t)','y(0)=0','Dy(0)=0','t');
>> y=simple(y);
```

The expression for the solution becomes somewhat simpler if we introduce the natural frequency of the spring, which is $\omega_0 = \sqrt{k/m}$. We do this by substituting $\omega_0^2 m$ for k.

```
>> y=subs(y,'w0*w0*m','k');
>> pretty(y)
```

```
            A (cos(w t) - cos(w0 t))
            ------------------------
                  2         2
                w0  m  -  w   m
```

Let's see what this looks like. Let's set $A = m = 1$, $\omega_0 = 10$, and $\omega = 12$. Then

```
>> ya=subs(y,1,'A'); ya=subs(ya,1,'m');
```

106

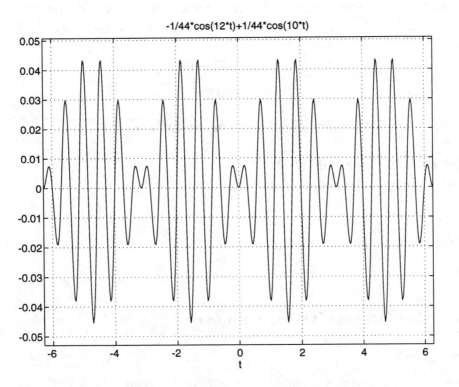

Figure 7.1. Beats in forced harmonic motion.

```
>> ya=subs(ya,12,'w'); ya=subs(ya,10,'w0')

ya =

-1/44*cos(12*t)+1/44*cos(10*t)

>> ezplot(ya)
```

produces Figure 7.1.

The graph in Figure 7.1 displays two oscillations. It shows a rapid, high frequency, oscillation with an amplitude which is oscillating at a smaller frequency. The formula plotted, and shown in the title of Figure 7.1, does not predict this result. This phenomenon is called *beats*. It occurs whenever two waves of slightly differing frequencies are allowed to interact. (In this case the frequencies are 10 and 12.) Beats are used by piano tuners when they tune a piano. One frequency is provided by a tuning fork, and the other by the vibrating piano string. The beats, i.e., the amplitude changing at the smaller frequency, can be heard, and the tuner tightens or loosens the piano string until the beats go away. At that point, the piano string will be vibrating with the same frequency as the tuning fork.

107

The phenomenon of beats becomes clearer after a little algebra. We introduce the mean of the two frequencies $\overline{\omega} = (\omega + \omega_0)/2$, and the half difference $\delta = (\omega - \omega_0)/2$. Then

$$\omega = \overline{\omega} + \delta$$
$$\omega_0 = \overline{\omega} - \delta$$

When we make these substitutions into the expression for y we get

```
>> y = subs(y,'wbar+del','w');
>> y = subs(y,'wbar-del','w0');
```

```
>> y=simple(y)

y =

1/2*A/m/wbar/del*sin(t*wbar)*sin(t*del)

>> pretty(y)
```

```
            A sin(t wbar) sin(t del)
      1/2   --------------------------
                    m wbar del
```

In a nicer form this is

$$y(t) = \frac{A \sin(\delta t)}{2\, m\overline{\omega}\, \delta} \sin(\overline{\omega} t). \tag{2}$$

From (2) we can understand beats better. The large frequency is $\overline{\omega}$ and the small frequency is δ. If we put our emphasis on the fast oscillation with frequency $\overline{\omega}$, then the amplitude is

$$\frac{A \sin(\delta t)}{2\, m\overline{\omega}\, \delta},$$

which is oscillating itself, but at a much smaller frequency.

A piano tuner, listening to a mistuned piano string and a tuning fork, hears both the large frequency, which is close to the required frequency, and the small frequency, which represents the error in the tuning. As the string is tuned the frequency δ gets smaller and ultimately the tone with that frequency becomes inaudible, at which point the string frequency is very close to the tuning fork frequency.

It is interesting to examine graphically what happens to (2) as $\delta \to 0$, i.e. as $\omega \to \omega_0$. We will leave this for the exercises. At this point it is sufficient to notice that if we set $\omega = \omega_0$ in the original equation, and then let MATLAB solve the equation we get:

```
>> y = dsolve('m*D2y+m*w*w*y=A*cos(w*t)','y(0)=0','Dy(0)=0','t');
>> pretty(y)
```

$$
\frac{1}{2} \; \frac{A \; \sin(w \; t) \; t}{w \; m}
$$

This is

$$
y(t) = \frac{At}{2\omega m} \sin(\omega t). \tag{3}
$$

Now the amplitude of the oscillation is $At/2\omega m$, which is steadily increasing with time. This is an illustration of resonance. We will see more of this in the next section.

The forced damped vibrating spring

Now we will consider equation (1) with nonzero damping. If we proceed as we did in the previous section by directly solving the initial value problem, we are led to expressions that are not too meaningful, and do not easily lead to understanding. Instead, we will use our knowledge of differential equations and proceed more indirectly.

We know that the general solution of (1) is of the form

$$
C_1 e^{r_1 t} + C_2 e^{r_2 t} + R \cos(\omega t - \delta), \tag{4}
$$

where C_1 and C_2 are arbitrary constants, and r_1 and r_2 are solutions of the characteristic equation

$$
mr^2 + dr + k = 0.
$$

This is a simple quadratic equation, so we can solve it without MATLAB's help. The solutions are

$$
r = \frac{1}{2m} \left(-d \pm \sqrt{d^2 - 4km} \right).
$$

Upon examination, we see that, since m, d, and k are all positive numbers, both solutions are either negative or at least have negative real part. As a result, the exponential part of (4) decays to 0 as time increases, and we are left with what is called the *steady state solution*:

$$
y_s = R \cos(\omega t - \delta). \tag{5}
$$

The amplitude R, and the phase δ in the steady state solution have yet to be determined. Let's use MATLAB to find them. We will substitute y_s into the differential equation using symop.

109

This substitution is too complicated to do in one step, so we will split the process into two.

```
>> ys = 'R*cos(w*t-del)';
>> eq = symop('m*',diff(ys,'t',2),'+','d*',diff(ys,'t'));
>> eq = symop(eq,'+','k*',ys,'=','A*cos(w*t)');
>> eq = simple(eq)

eq =

R*(-m*cos(-w*t+del)*w^2+d*sin(-w*t+del)*w+k*cos(-w*t+del)) = A*cos(w*t
```

We must find R and δ so that this equation is true for all values of t. If this is true, then the equation and its derivative will both be valid for $t = 0$. We will make these two substitutions to get two equations which do not have t as a variable.

```
>> eq1=subs(eq,0,'t');
>> eq2=subs(diff(eq,'t'),0,'t');
```

You can display these equations if you wish.

Next we solve these equations for R and δ.

```
>> [R,del] = solve(eq1,eq2,'R,del')

R =

A/(k^2-2*k*w^2*m+w^4*m^2+d^2*w^2)^(1/2)



del =

atan(d*w/(k-w^2*m))
```

We are particularly interested in the amplitude R, and how it varies with the driving frequency ω. The results will look better if we substitute the natural frequency of the corresponding undamped system $\omega_0 = \sqrt{k/m}$.

```
>> R = subs(R,'w0*w0*m','k')

R =

A/(w0^4*m^2-2*w0^2*m^2*w^2+w^4*m^2+d^2*w^2)^(1/2)
```

If we use `pretty(R)` and a little of our own algebra, we see that

$$R = \frac{A}{\sqrt{m^2(\omega^2 - \omega_0^2)^2 + d^2\omega^2}}. \qquad (6)$$

From (6) we see that $R \to 0$ as $\omega \to \infty$, and that when $\omega = 0$, $R = A/m\omega_0^2$. It is also clear that R achieves its maximum value at ω somewhere near ω_0. To find the maximum point exactly, we use MATLAB to find out where the derivative is equal to 0. We solve the equation $\frac{dR}{d\omega} = 0$ for ω.

```
>> wmax = solve(diff(R,'w'),'w')

wmax =

[-1/2*2^(1/2)/m*(2*w0^2*m^2-d^2)^(1/2)]
[-1/2*2^(1/2)/m*(2*w0^2*m^2-d^2)^(1/2)]
[-1/2*2^(1/2)/m*(2*w0^2*m^2-d^2)^(1/2)]
[ 1/2*2^(1/2)/m*(2*w0^2*m^2-d^2)^(1/2)]
```

MATLAB has found four solutions. Three of them are equal and negative, but we want the positive one, so we use `sym` to pick it out.

```
>> wmax = sym(wmax,4,1);
>> pretty(wmax)

                  1/2        2  2      2 1/2
                 2      (2 w0  m   - d )
           1/2   -----------------------
                           m
```

Thus we see that the maximum is achieved when

$$\omega^2 = \omega_{max}^2 = \omega_0^2 - \frac{d^2}{2m^2}.$$

To find the maximum value of R, we substitute ω_{max} into the expression for R.

```
>> Rmax = subs(R,wmax,'w');
```

111

```
>> Rmax = simple(Rmax);
>> pretty(Rmax)
```

```
                                      A
           ---------------------------------------
           /                          4 \1/2
           |   2   2                  d  |
           |w0   d   - 1/4  ----|
           |                          2  |
           \                        m    /
```

With a little of our own algebra, we see that

$$R_{\max} = \frac{A}{d\sqrt{\omega_0^2 - d^2/4m^2}}.$$

Finally we would like to plot R. Currently there are too many variables in the expression for R. We will lump constants to reduce the number without losing the physical significance. First it is natural to write the formula in terms of the ratio $s = \omega/\omega_0$. To do that we substitute $\omega = s\omega_0$.

```
>> R=subs(R,'s*w0','w');
>> pretty(R)
```

```
                                          A
      -----------------------------------------------------------------
        4  2        4  2  2    4   4  2     2  2    2 1/2
      (w0   m  - 2 w0   m  s  + s  w0   m  + d  s  w0 )
```

We notice that every term in the denominator has the factor $\omega_0^4 m^2$ except the last term. We can arrange for the last term to have this factor by introducing the lumped constant $D = d/m\omega_0$. We do this by substituting $d = Dm\omega_0$.

```
>> R = subs(R,'D*m*w0','d');
>> pretty(R)
```

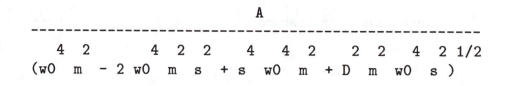

```
                                          A
      -----------------------------------------------------------------
        4  2        4  2  2    4   4  2     2  2    4  2 1/2
      (w0   m  - 2 w0   m  s  + s  w0   m  + D  m  w0   s )
```

112

Finally we choose the amplitude of the driving force $A = m\omega_0^2$.

```
>> R=subs(R,'m*w0*w0','A');
>> R = simple(R);
>> pretty(R)
```

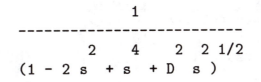

Amplitude of the steady state response vs. frequency.

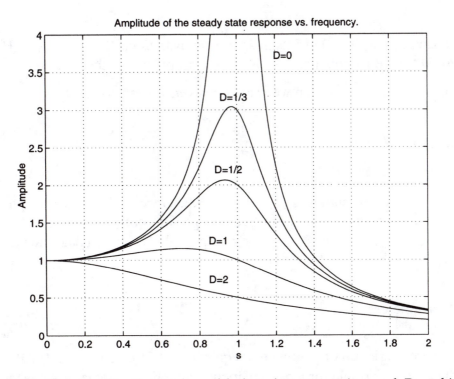

Figure 7.2. Forced, vibrating spring with damping. $s = \omega/\omega_0$, and $D = d/m\omega_0$

We recognize this as

$$R = \frac{1}{\sqrt{(1 - s^2)^2 + D^2 s^2}}. \tag{7}$$

Thus we have expressed the amplitude of the steady state oscillation as a function of $s = \omega/\omega_0$, with the single parameter $D = d/m\omega_0$. Notice that D is proportional to d, and consequently represents the damping force.

To plot R, we can use `ezplot` as we did in a previous section. It is necessary to use `hold on`, and to substitute the value of the parameter D each time.

```
>> ezplot(subs(R,1,'D'),[0,2])
>> hold on
>> ezplot(subs(R,2,'D'),[0,2])
>> ezplot(subs(R,1/2,'D'),[0,2])
>> ezplot(subs(R,1/3,'D'),[0,2])
>> ezplot(subs(R,0,'D'),[0,2])

Warning: Divide by zero
>> hold off
```

The computation for $D = 0$ results in a division by 0, as can be seen in (7). However, this does not cause MATLAB to stop, nor does it cause a problem with the plot.

The scale of the plot is not what we want, however, we can correct this with

```
>> axis([0,2,0,4])
```

The result, after a little sprucing up, is Figure 7.2. Once again we see the phenomenon of resonance. Now we see that oscillations of very large amplitude can result if there is not very much damping in the system and if the driving frequency ω is chosen near the maximum ω_{max}.

Variation of parameters

The method of variation of parameters requires a lot of algebra. MATLAB is quite good at deriving the general formula for a particular equation. For example, if we want to solve the inhomogeneous equation

$$y'' - 3y' + 2y = g(t)$$

using variation of parameters, the standard way requires us to find two linearly independent solutions to the homogeneous equation, and then to do a lot of algebra before we can find the formula for the solution. Using MATLAB we can simply proceed as follows:

```
>> eq = 'D2y-3*Dy+2*y=g(t)';
>> y = dsolve(eq,'t');
```

114

```
>> pretty(y)
```

```
    /                        /
    |                        |
    |  - exp(- t) g(t) dt exp(t) +  |  exp(- 2 t) g(t) dt exp(2 t)
    |                        |
    /                        /


        + C1 exp(t) + C2 exp(2 t)
```

This is easily recognizable as

$$y(t) = \int e^{-2t} g(t) dt \, e^{2t} - \int e^{-t} g(t) dt \, e^{t} + C1 \, e^{2t} + C2 \, e^{t}.$$

This gives us the solution in complete generality, including the general solution to the homogeneous equation. If we want to find the solution with initial conditions $y(0) = 1$, and $y'(0) = 0$, it is just as easy:

```
>> y =  dsolve(eq,'y(0)=1','Dy(0)=0','t');
>> pretty(y)
```

```
    t                        t
    /                        /
    |                        |
    |  - exp(- u) g(u) du exp(t) +  |  exp(- 2 u) g(u) du exp(2 t)
    |                        |
    /                        /
    0                        0


        + 2 exp(t) - exp(2 t)
```

This will work for equations with variable coefficients, provided that MATLAB is able to find the solutions to the homogeneous equation. For example, in the case of the equation

$$y'' - \frac{2t}{1 + t^2} y' + \frac{2}{1 + t^2} y = g(t),$$

if we simply give MATLAB a chance we get an answer.

115

```
>> eq = 'D2y-2*t*Dy/(1+t^2)+2*y/(1+t^2) = g(t)';
>> y = dsolve(eq,'t');
>> pretty(y)
```

```
  /                    /                 /             2
 |   t g(t)       2   |   t g(t)        |    g(t) (t  - 1)
 |   ------ dt t   -  |   ------ dt  +  |  - ------------- dt t
 |      2            |      2          |          2
 /   1 + t           /   1 + t         /       1 + t

              2
     + C1 t  - C1 + C2 t
```

Systems of equations

MATLAB with dsolve can solve some systems of differential equations. For example, for the simple system

$$x' = y$$
$$y' = -x$$

we use

```
>> sys = 'Dx=y,Dy=-x'

sys =

Dx=y,Dy=-x
```

Then solving the system is no more difficult than before.

```
>> [x,y]=dsolve(sys)

x =

C2*sin(t)-C1*cos(t)


y =

C1*sin(t)+C2*cos(t)
```

116

MATLAB can also solve the slightly more complicated system

$$x' = y$$
$$y' = -x + 0.1\,y$$

with initial conditions $x(0) = 1$, $y(0) = 0$.

```
>> sys = 'Dx=y,Dy=-x+0.1*y'

sys =

Dx=y,Dy=-x+0.1*y

>> init='x(0)=1,y(0)=0'

init =

x(0)=1,y(0)=0
```

Then

```
>> [x,y] = dsolve(sys,init);
```

solves the equations. (The solutions are not printed here because they are somewhat complicated.)

Interestingly, the solutions to this system have a different appearance if we enter the system in a more algebraic way.

```
>> sys = 'Dx=y,Dy=-x+y/10'

sys =

Dx=y,Dy=-x+y/10
```

This time the answer is

```
>> [x,y]=dsolve(sys,init);
```

117

```
>> pretty(x)
```

$$
\begin{aligned}
&\quad\quad\quad\ \ 1/2\ \ 1/2\ \ \ 1/2 \\
&- 1/399\ 3\quad 7\quad\ \ 19 \quad \exp(1/20\ t)\ \sin(1/20\ 3 \overset{1/2}{\ }\ 7 \overset{1/2}{\ }\ 19 \overset{1/2}{\ } \\
\\
&\quad\quad\quad\quad\quad\quad\quad\quad\quad\quad\ \ 1/2\ \ 1/2\ \ 1/2 \\
&+ \exp(1/20\ t)\ \cos(1/20\ 3 \quad 7 \quad 19 \quad\ t)
\end{aligned}
$$

```
>> pretty(y)
```

$$
\begin{aligned}
&\quad\quad 20\ \ 1/2\ \ 1/2\ \ \ 1/2 \\
&- \text{---}\ 3\quad 7\quad\ \ 19 \quad\ \ \exp(1/20\ t)\ \sin(1/20\ 3 \overset{1/2}{\ }\ 7 \overset{1/2}{\ }\ \ 19 \overset{1/2}{\ }\ \ t) \\
&\ \ 399
\end{aligned}
$$

The results can be plotted in several ways. However, `ezplot` can only be used to plot one of the components of the solution at a time, e.g. `ezplot(x,[0,10])` or `ezplot(y,[0,10])`. To see both components of the solution plotted on the same figure, we use `hold on` as before.

```
>> ezplot(x,[0,10])
>> hold on
>> ezplot(y,[0,10])
>> hold off
```

will do the trick.

The phase plane plot of y vs. x requires the new commands `vectorize` and `eval`. The command `vectorize` makes a symbolic expression array smart, so that it can be evaluated on a matrix. Look at what happens when you execute `xv = vectorize(x)`. The command `eval` evaluates the string which is its parameter. Notice that the symbolic expression `xv` has a `t` in it. If we define `t=0:0.1:10;`, then we can evaluate `xv` with the command `eval(xv)`. Putting this all together, we can get the phase plane plot with the following sequence of commands:

```
>> xv = vectorize(x);
>> yv = vectorize(y);
>> t=0:0.1:10;
>> plot(eval(xv),eval(yv))
```

Let's try it on something a little harder:

$$
\begin{aligned}
x' &= 3x + 2y + 4z \\
y' &= 2x + 2z \\
z' &= 4x + 2y + 3z
\end{aligned}
$$

First, we enter the system.

```
>> sys = 'Dx = 3*x+2*y+4*z, Dy = 2*x+2*z, Dz = 4*x+2*y+3*z'

sys =

Dx = 3*x+2*y+4*z, Dy = 2*x+2*z, Dz = 4*x+2*y+3*z
```

Then we solve it.

```
>> [x,y,z] = dsolve(sys)

x =

C2*exp(8*t)+C3*exp(-t)


y =

1/2*C2*exp(8*t)+(-2*C1-2*C3)*exp(-t)


z =

C2*exp(8*t)+exp(-t)*C1
```

Warning! It is tempting to write systems of differential equations using variables with subscripts; for example, $x'_1 = x_2$, $x'_2 = -x_1$. It seems natural to extend the subscript usage to MATLAB, using `x1` for x_1, etc. However, MATLAB does not accept this usage. To see what happens, try `dsolve('Dx1=0')`.

Exercises

1. Find the general solution to each of the following differential equations:
   a) $y'' - 3y' + 2y = 0$.
   b) $y'' - 3y' + 8y = 0$.
   c) $y'' - 4y' + 5y = 0$.
   d) $y'' - 4y' + 5y = e^{-t}$.
   e) $y'' - 4y' + 5y = \cos(t)$.
   f) $y'' - 4y' + 5y = e^{-t}\cos(t)$.

119

2. Find the solution to each of the differential equations in Exercise 1 with the initial conditions $y(0) = 0$, and $y'(0) = 1$.

3. Find the general solution to each of the following differential equations:

   a) $t^2 y'' = (y')^2$.

   b) $y'' = y y'$.

   c) $y'' = t y' + y + 1$. You will have to learn about the error function. A quick way is to execute `help erf`.

4. Find the solutions to the following initial value problems:

   a) $t^2 y'' = (y')^2$, with $y(1) = 3$ and $y'(1) = 2$.

   b) $y'' = y y'$, with $y(0) = 0$, and $y'(0) = 2$. MATLAB may need help with this one.

   c) $y'' = t y' + y + 1$, with $y(0) = 1$, and $y'(0) = 0$.

5. The purpose of this exercise is to examine the behavior of undamped, forced harmonic motion as the driving frequency approaches the natural frequency of the spring. Thus, we will be examining formula (2) as $\delta \to 0$, and comparing the behavior to the case when the driving frequency is equal to the natural frequency, which is given by (3). For convenience, set $\overline{\omega} = 1$ in these formulas, and assume that $A = 2m\omega$.

   For $\delta = 2$, 1, 0.5, 0.2, and 0.1, plot the function in (2) and the function in (3) over the interval $[0, 4\pi]$. Superimpose the two graphs on the same figure. Describe what you see happening as $\delta \to 0$.

8.   Linear Algebra Using MATLAB

MATLAB is short for "Matrix Laboratory". It contains a large number of built-in routines that make matrix algebra very easy. In this chapter we will learn how to use some of those routines to solve systems of linear equations.

Systems of linear equations

We are interested in solving systems of equations like

$$u + 2v = 5$$
$$4u - v = 2. \tag{1}$$

This is a system of two equations involving two unknowns. We want to find a systematic method that will work in general for m equations involving n unknowns, where m, and n are possibly very large, and not necessarily equal.

The general method is not significantly different from the method usually taught in secondary school to solve systems like (1). We solve the first equation for the variable u, yielding $u = 5 - 2v$, and then we substitute for u in the second equation. In this way we obtain a new system consisting of the first equation and the modified second equation

$$u + 2v = 5$$
$$-9v = -18, \tag{2}$$

which has the same solutions as (1). The advantage is that system (2) is quite easily solved. From the second equation, we see that $v = 2$. Then substituting $v = 2$ into the first equation, we can solve for u, getting $u = 1$.

Solving and substituting gets the job done, but this is difficult to implement for larger systems. The key to developing a more systematic approach is to notice that we can get the system in (2) by subtracting 4 times the first equation in (1) from the second equation in (1). This operation of adding (or subtracting) a multiple of one equation to (or from) another is equivalent to solving and substituting, but is more direct and easier to implement.

While the operation of adding a multiple of one equation of a system to another has been illustrated only for the special case of (1), it is easily seen that such an operation on an arbitrary system leads to a new system which is equivalent to the first in the sense that any solution of one system is also a solution of the other.

Our next step in devising a general solution method is to get rid of some excess baggage. We look at the matrix equation for the system. In the case of (1), we let

$$A = \begin{pmatrix} 1 & 2 \\ 4 & -1 \end{pmatrix}, \quad x = \begin{pmatrix} u \\ v \end{pmatrix}, \quad \text{and} \quad b = \begin{pmatrix} 5 \\ 2 \end{pmatrix}.$$

Then the matrix equation $\mathbf{Ax} = \mathbf{b}$ is equivalent to the system of equations in (1). All of the information about the system is contained in the *coefficient matrix* $\mathbf{A}$, and the *right hand side* $\mathbf{b}$. From these two we form the *augmented matrix* $\mathbf{a}$ by adding the column vector $\mathbf{b}$ as a third column to the matrix $\mathbf{A}$; i.e.,

$$\mathbf{a} = [\mathbf{A}, \mathbf{b}] = \begin{pmatrix} 1 & 2 & 5 \\ 4 & -1 & 2 \end{pmatrix}.$$

The augmented matrix contains all of the information about the system in a compact form.

Notice that each row in $\mathbf{a}$ corresponds to one of the equations in (1). Furthermore, if we add -4 times the first row of $\mathbf{a}$ to the second row, we get

$$\begin{pmatrix} 1 & 2 & 5 \\ 0 & -9 & -18 \end{pmatrix},$$

which is the augmented matrix corresponding to the system in (2). Thus we see that the method of solving and substitution in the system of equations becomes the operation of adding a multiple of one row of the augmented matrix to another row.

This is our first example of a *row operation* on a matrix. There are three in all, and they are:

R1. Add a multiple of one row in a matrix to another row.
R2. Interchange two rows of the matrix.
R3. Multiply a row of a matrix by a nonzero constant.

Notice that if the matrix is the augmented matrix associated to a system of equations, then operation R2 corresponds to interchanging two of the equations, and R3 corresponds to multiplying an equation by a nonzero constant. Thus each of these two row operations transforms the augmented matrix into the augmented matrix of another system of equations which has exactly the same solutions as the original. We have already seen that the first row operation, R1, also has this property.

Our general strategy for solving systems is to replace the system by the augmented matrix, and then perform row operations until the system associated with the transformed matrix is easy to solve. Of course, each of the row operations requires a significant amount of arithmetic, but we will let MATLAB do that arithmetic for us. We will discuss how to do this in MATLAB after addressing some preliminaries.

Matrix indexing in MATLAB and row operations

One of the key features of MATLAB is the ease with which the elements of a matrix can be accessed and manipulated. We will explain some of these features here, in preparation for our explanation of how to implement row operations on matrices.

If M is a matrix in MATLAB, then the entry in the second row and fourth column is denoted by M(2,4). Thus, for example:

```
>> M=[1 3 5 6;3 5 0 -3;-4 0 9 3]

M =

     1     3     5     6
     3     5     0    -3
    -4     0     9     3

>> M(2,4)

ans =

    -3
```

Try M(1,3).

The entries in a matrix can be changed easily. For example, if we want to change M(3,4) to be equal to -15, we enter

```
>> M(3,4)=-15


M =

     1     3     5     6
     3     5     0    -3
    -4     0     9   -15
```

and the job is done.

We can refer to the second row in the matrix M by M(2,:). The colon means we want all elements of the particular row:

```
>> M(2,:)

ans =

     3     5     0    -3
```

Notice also that >> M(:,2) displays the second column of M. The basic idea is that the first index refers to the row, and the second to the column.

We can also easily refer to submatrices. For example, if we want to refer to the matrix consisting of the first and third rows of M, we enter

```
>> M([1 3],:)
```

```
ans =
```

```
    1      3      5      6
   -4      0      9    -15
```

and we get what we want. Of course, we can do the same with columns:

```
>> M(:,[2,4])
```

```
ans =
```

```
    3      6
    5     -3
    0    -15
```

What do you think M([1,3],[4,1]) refers to? Try it and see.

Now we are ready to explain how to do row operations. We will start with the matrix **M** we ended up with above, i.e.

```
>> M
```

```
M =
```

```
    1      3      5      6
    3      5      0     -3
   -4      0      9    -15
```

We will illustrate the operation R1 by adding 4 times the first row to the third row. The notation for the third row is M(3,:), so we will replace this by M(3,:) + 4*M(1,:). We will let MATLAB do all of the tedious arithmetic.

```
>> M(3,:) = M(3,:)+4*M(1,:)
```

```
M =
```

```
    1      3      5      6
    3      5      0     -3
    0     12     29      9
```

124

To illustrate operation R2 we will exchange rows 2 and 3. In this case we want `M([2,3],:)` to be equal to what `M([3,2],:)` is currently, so:

```
>> M([2,3],:)=M([3,2],:)

M =

    1     3     5     6
    0    12    29     9
    3     5     0    -3
```

Finally, we illustrate R3 by multiplying row 2 by -5.

```
>> M(2,:) = -5*M(2,:)

M =

    1     3      5      6
    0   -60   -145    -45
    3     5      0     -3
```

We can divide a row by a number just as easily

```
>> M(3,:) = M(3,:)/3

M =

   1.0000    3.0000     5.0000     6.0000
        0  -60.0000  -145.0000   -45.0000
   1.0000    1.6667          0    -1.0000
```

The rational format

Frequently problems arise in which the numbers we are using are all rational numbers. The floating point arithmetic that MATLAB uses does not lend itself to maintaining the rational form of numbers. We can sometimes recognize a rational number from the decimal expansions MATLAB gives us, but frequently, we can't. For example, in M as it appears above, we know that `M(3,2)=5/3`, but we will not always be so fortunate. MATLAB provides a solution with its rational format. If we enter `format rat` at the MATLAB prompt, from that point on MATLAB will display numbers as the ratios of integers. For example,

```
>> format rat
>> pi

ans =

    355/113
```

We know, and MATLAB knows, that $\pi \neq 355/113$. MATLAB is only displaying an approximation to the answer. But this is no different than what MATLAB always does. The only difference is that now Matlab is using a rational number to approximate π instead of a decimal expansion. It computes and displays the rational number that is closest to the given number within a prescribed tolerance. The advantage is that if we know the answer is a rational number, then we know that the rational representation shown us by MATLAB is probably 100% correct.

If we display our matrix M now, we see the following:

```
>> M

M =

    1           3           5           6
    0          -60        -145        -45
    1          5/3         0           -1
```

Thus we get the precise, rational version of M.

There are some things about the rational format that can be confusing. We will inevitably run into them as we proceed. For one thing, asterisks sometimes appear in the output. These appear when MATLAB is trying to give a rational approximation to a very small number. This will require a very large denominator, and the resulting expression will not fit within the rational format. When we are dealing with what we know to be rational numbers with small denominators, these large denominators that appear are due to round off error. They usually represent a number which actually should be 0, and they can usually be replaced by 0, but caution is called for in doing so.

Solving linear equations

We are ready to demonstrate how to use row operations to solve systems of equations. Consider the system

$$3x_1 - 4x_2 + 5x_3 - x_4 = 5$$
$$x_2 - x_3 + 6x_4 = 0$$
$$5x_1 - x_3 + 4x_4 = 4$$

126

In terms of the coefficient matrix

$$M = \begin{pmatrix} 3 & -4 & 5 & -1 \\ 0 & 1 & -1 & 6 \\ 5 & 0 & -1 & 4 \end{pmatrix}$$

and the vectors

$$\mathbf{x} = \begin{pmatrix} x_1 \\ x_2 \\ x_3 \\ x_4 \end{pmatrix} \qquad \mathbf{b} = \begin{pmatrix} 5 \\ 0 \\ 4 \end{pmatrix},$$

the system of equations is equivalent to the equation

$$\mathbf{Mx} = \mathbf{b}.$$

In MATLAB, we proceed as follows:

```
>> M=[3 -4 5 -1;0 1 -1 6;5 0 -1 4]

M =

     3        -4         5        -1
     0         1        -1         6
     5         0        -1         4

>> b=[5 0 4]'

b =

     5
     0
     4
```

In this last command, the prime (') tells MATLAB to take the *transpose* of the given matrix. In general this flips the matrix along its diagonal. In the case of a vector, it changes a row vector to a column vector and vice versa.

To analyze the system using row operations, we augment the matrix **M** by adding the vector **b** as a fifth column. We will denote the augmented matrix by **m**.

```
>> m=[M,b]

m =
```

3	-4	5	-1	5
0	1	-1	6	0
5	0	-1	4	4

Now we transform **m** by row operations to get an equivalent system which is easier to solve. We start by making $m(1,1)=1$. To do this we divide row 1 by 3:

```
>> m(1,:)=m(1,:)/3

m =
```

1	-4/3	5/3	-1/3	5/3
0	1	-1	6	0
5	0	-1	4	4

Notice that we are using the rational format.

Next we eliminate the non-zero entries in the first column by adding or subtracting appropriate multiples of the first row.

```
>> m(3,:)=m(3,:)-5*m(1,:)

m =
```

1	-4/3	5/3	-1/3	5/3
0	1	-1	6	0
0	20/3	-28/3	17/3	-13/3

Then we eliminate the entries in column 2 except $m(2,2)$ by adding multiples of the second row to the first and the third.

```
>> m(1,:)=m(1,:)+(4/3)*m(2,:)
```

128

```
m =
```

1	0	1/3	23/3	5/3
0	1	-1	6	0
0	20/3	-28/3	17/3	-13/3

```
>> m(3,:)=m(3,:)-(20/3)*m(2,:)

m =
```

1	0	1/3	23/3	5/3
0	1	-1	6	0
0	*	-8/3	-103/3	-13/3

That asterisk (*) in the bottom row should be a zero. To find out what it is exactly, enter

```
>> m(3,2)

ans =

    -1/1125899906842624
```

Thus the asterisk indicates that the rational approximation of the number is too large to fit into the format of the matrix. But m(3,2) should be precisely zero, not almost zero as it is here. We have been a little careless. When we do row operations, we should refer to the matrix elements by name instead of trying to guess the number that MATLAB is using for them. Remember, MATLAB is dealing with approximations all the time. It is constantly making very small mistakes due to round off error. To fix this we proceed:

```
>> m(3,:)=m(3,:)-m(3,2)*m(2,:)

m =
```

1	0	1/3	23/3	5/3
0	1	-1	6	0
0	0	-8/3	-103/3	-13/3

Notice that we referred directly to m(3,2) in this command. This is the preferred method for doing row operations. We will use this procedure in the future, and we recommend that you do too.

Next we divide the third row by m(3,3) to make that element 1.

```
>>  m(3,:)=m(3,:)/m(3,3)

m =
```

1	0	1/3	23/3	5/3
0	1	-1	6	0
0	0	1	103/8	13/8

Then we eliminate in the third column by adding appropriate multiples of the third row.

```
>> m(1,:)=m(1,:)-m(1,3)*m(3,:);
>> m(2,:)=m(2,:)-m(2,3)*m(3,:)

m =
```

1	0	0	27/8	9/8
0	1	0	151/8	13/8
0	0	1	103/8	13/8

Since we started with all integer entries, we can be sure that all of these entries are rational numbers, so what we see here is the correct answer. We can see what MATLAB thinks the answer is in decimal notation by switching to `format long`

```
>> format long
>> m

m =

Columns 1 through 4

   1.00000000000000                  0                  0   3.37500000000
                  0   1.00000000000000                  0  18.87499999999
                  0                  0   1.00000000000000  12.87499999999

Column 5

   1.12500000000000
   1.62500000000000
   1.62500000000000
```

130

Thus m(3,4) is equal to 103/8 in the rational format and 12.87499999999999 in the long format. Both representations are close to the correct answer, but, because we know that the answer is a rational number, we can be sure that 103/8 is correct.

At any rate, we see that the original system of equations is equivalent to the system

$$x_1 + \frac{27}{8}x_4 = \frac{9}{8}$$
$$x_2 + \frac{151}{8}x_4 = \frac{13}{8}$$
$$x_3 + \frac{103}{8}x_4 = \frac{13}{8}.$$

In this system we can assign any value to x_4 and then solve for the other unknowns. x_4 is an example of a *free variable*. If we set $x_4 = t$, then the general solution is

$$x_1 = (9 - 27t)/8$$
$$x_2 = (13 - 151t)/8$$
$$x_3 = (13 - 103t)/8$$
$$x_4 = t$$

where t is arbitrary.

Let's do another example quickly. We will find the general solution to the system of equations $\mathbf{M}\mathbf{x} = \mathbf{b}$, where

$$\mathbf{M} = \begin{pmatrix} 2 & -2 & 3 & 4 \\ -3 & 0 & 0 & 0 \\ 2 & 7 & 9 & 0 \end{pmatrix}, \qquad \mathbf{b} = \begin{pmatrix} 0 \\ 1 \\ 9 \end{pmatrix}$$

In MATLAB we enter

```
>> M=[2 -2 3 4;-3 3 0 0;2 7 9 0];
>> b=[0 1 9]';
```

We augment **M**:

```
>> m=[M,b]

m =

     2    -2     3     4     0
    -3     3     0     0     1
     2     7     9     0     9
```

131

We make $m_{11} = 1$:

```
>> m(1,:)=m(1,:)/m(1,1);
```

We eliminate the nonzero entries in the first column:

```
>> m(2,:)=m(2,:)-m(2,1)*m(1,:);
>> m(3,:)=m(3,:)-m(3,1)*m(1,:)
m =
```

1	-1	3/2	2	0
0	0	9/2	6	1
0	9	6	-4	9

We want m(2,2) to be nonzero. To effect this we use R2 and interchange the second and third rows. Then we make m(2,2) equal to 1, and finally eliminate the nonzero entries in the second column.

```
>> m([2,3],:)=m([3,2],:);
>> m(2,:)=m(2,:)/m(2,2);
>> m(1,:)=m(1,:)-m(1,2)*m(2,:)


m =
```

1	0	13/6	14/9	1
0	1	2/3	-4/9	1
0	0	9/2	6	1

We go through a similar procedure with the third column:

```
>> m(3,:)=m(3,:)/m(3,3);
>> m(1,:)=m(1,:)-m(1,3)*m(3,:);
>> m(2,:)=m(2,:)-m(2,3)*m(3,:)


m =
```

1	0	0	-4/3	14/27
0	1	0	-4/3	23/27
0	0	1	4/3	2/9

Thus the original system of equations is equivalent to the system:

$$x_1 - (4/3)x_4 = 14/27$$
$$x_2 - (4/3)x_4 = 23/27$$
$$x_3 + (4/3)x_4 = 2/9$$

We see that x_4 is a free variable, so we set $x_4 = s$. Then the complete, general solution is

$$\mathbf{x} = \begin{pmatrix} 14/27 + (4/3)s \\ 23/27 + (4/3)s \\ 2/9 - (4/3)s \\ s \end{pmatrix} = \begin{pmatrix} 14/27 \\ 23/27 \\ 2/9 \\ 0 \end{pmatrix} + s \begin{pmatrix} 4/3 \\ 4/3 \\ -4/3 \\ 1 \end{pmatrix},$$

where s is arbitrary.

Reduced row echelon form

In each of the examples of the previous section, the last form of the matrix m is called the *reduced row echelon form* of the original matrix. In general a matrix is said to be in reduced row echelon form if it looks like (3), where the asterisks stand for arbitrary numbers.

$$\begin{pmatrix} 1 & 0 & * & * & 0 & 0 & * & * & 0 & * \\ 0 & 1 & * & * & 0 & 0 & * & * & 0 & * \\ 0 & 0 & 0 & 0 & 1 & 0 & * & * & 0 & * \\ 0 & 0 & 0 & 0 & 0 & 1 & * & * & 0 & * \\ 0 & 0 & 0 & 0 & 0 & 0 & 0 & 0 & 1 & * \\ 0 & 0 & 0 & 0 & 0 & 0 & 0 & 0 & 0 & 0 \\ 0 & 0 & 0 & 0 & 0 & 0 & 0 & 0 & 0 & 0 \end{pmatrix}. \tag{3}$$

A reduced row echelon matrix has these main features:

- Each nonzero row starts with (≥ 0) elements which are equal to zero, followed by a 1, which is called the *pivot* of the row.
- The pivot in each row is strictly to the right of the pivot in the row above.
- There may be rows consisting entirely of zeros, but these rows are collected at the bottom of the matrix.
- A column which contains a pivot is called a *pivot column*, and the pivot is the only nonzero element in a pivot column.
- A column which does not contain a pivot is called a *free column*.

Notice that the system associated with a matrix which is in reduced row echelon form is extremely easy to solve. Each variable corresponds to a column in the matrix. It is a *free variable* if the column is a free column, and a *pivot variable* if the column is a pivot column. The value of the free variables can be assigned arbitrarily. Then each row of the matrix corresponds to an equation which expresses one of the pivot variables in terms of the right-hand side and the free variables. It is usually possible to write down the solutions by observation as we did with the examples in the previous section.

Our entire strategy for solving systems of equations amounts to taking the augmented matrix of the system and transforming it to reduced row echelon form using row operations. The method is called *Gaussian elimination*.

After doing these examples, the reader must be wondering why the computer and MATLAB can't do the whole process at once. In fact it can. If we remember **M** and **b** from the last example, and we form the augmented matrix **m**

```
>> m=[M,b]
```

m =

2	-2	3	4	0
-3	3	0	0	1
2	7	9	0	9

then we proceed directly to the reduced row echelon form by using the MATLAB command `rref`.

```
>> mr=rref(m)
```

mr =

1	0	0	-4/3	14/27
0	1	0	-4/3	23/27
0	0	1	4/3	2/9

This eliminates all of the intermediate steps, and we can immediately write down the equivalent system of equations as we did earlier.

Let's do one more example. We will solve the system

$$2x_2 + 2x_3 + 3x_4 = -4$$
$$-2x_1 + 4x_2 + 2x_3 - x_4 = -6$$
$$3x_1 - 4x_2 - x_3 + 2x_4 = 8.$$

The augmented matrix is

```
>> m = [0 2 2 3 -4; -2 4 2 -1 -6; 3 -4 -1 2 8]

m =

     0          2          2          3         -4
    -2          4          2         -1         -6
     3         -4         -1          2          8
```

Applying `rref` we get

```
>> rref(m)

ans =

     1          0          1          0        16/5
     0          1          1          0        -1/5
     0          0          0          1        -6/5
```

Thus we get the equivalent system

$$x_1 + x_3 = 16/5$$
$$x_2 + x_3 = -1/5$$
$$x_4 = -6/5$$

This time x_3 is a free variable. Solving as before, we get

$$\mathbf{x} = \begin{pmatrix} 16/5 - t \\ -1/5 - t \\ t \\ -6/5 \end{pmatrix} = \begin{pmatrix} 16/5 \\ -1/5 \\ 0 \\ -6/5 \end{pmatrix} + t \begin{pmatrix} -1 \\ -1 \\ 1 \\ 0 \end{pmatrix}.$$

Determined systems of equations

A system is *determined* if there are the same number of equations and unknowns. It is *underdetermined* if there are fewer equations than unknowns, and *overdetermined* if there are more. The systems we have been solving up to now have been underdetermined. There are special methods in MATLAB for studying determined systems, which we will discuss in this section.

For example, let

$$M_4 = \begin{pmatrix} 4 & -4 & -8 \\ 0 & 2 & 2 \\ 1 & -2 & -3 \end{pmatrix}, \quad M_3 = \begin{pmatrix} 3 & -4 & -8 \\ 0 & 2 & 2 \\ 1 & -2 & -3 \end{pmatrix}, \quad \text{and} \quad b = \begin{pmatrix} 27 \\ -6 \\ 10 \end{pmatrix}.$$

The subscript on the matrices refers to the entry in the first row and column. Notice that the other entries of M_3 and M_4 are equal.

Let's enter these matrices into MATLAB.

```
>> M4 = [4 -4 -8;0 2 2 ;1 -2 -3];
>> M3=M4;M3(1,1)=3

M3 =

    3    -4    -8
    0     2     2
    1    -2    -3
```

The augmented matrix for the system $M_4 x = b$ is

```
>> b =  [27;-6;10];m4=[M4,b]

m4 =

    4    -4    -8    27
    0     2     2    -6
    1    -2    -3    10
```

Applying rref we get

```
>> rref(m4)

ans =

    1     0    -1     0
    0     1     1     0
    0     0     0     1
```

Take notice of the last row in this matrix. It corresponds to the equation $0 = 1$, which has no solution. Since the system of equations corresponding to the reduced row echelon form has the same solutions as the original system, we conclude that the system $M_4 x = b$ has no solutions.

On the other hand, in the same way we discover that the homogeneous system $M_4x = 0$ is solved by any vector of the form

$$x = t \begin{pmatrix} 1 \\ -1 \\ 1 \end{pmatrix}.$$

(Verify this.)

If we look at the same equation with M_4 replaced by M_3, which differs only slightly, we get quite different behavior.

```
>> m3=[M3,b];
>> rref(m3)

ans =

    1    0    0    1
    0    1    0    0
    0    0    1   -3
```

Thus the system $M_3x = b$ has the unique solution

$$x = \begin{pmatrix} 1 \\ 0 \\ -3 \end{pmatrix}.$$

In the same way we discover that the homogeneous equation $Mx = 0$ has the unique solution 0.

These examples show that the situation with determined systems is not so simple. Sometimes there are solutions, sometimes not. Sometimes solutions are unique, sometimes not. We need a way to tell which phenomena occur in specific cases. The answers are provided by the *determinant*.

We will not study the determinant in any detail. Let's just recall some facts. The determinant is defined for any square matrix, i.e. any matrix which has the same number of rows and columns. For two by two and three by three matrices we have the formulas

$$\det \begin{pmatrix} a & b \\ c & d \end{pmatrix} = ad - bc$$

and

$$\det \begin{pmatrix} a & b & c \\ d & e & f \\ g & h & i \end{pmatrix} = aei - afh - bdi + cdh + bfg - ceg.$$

For larger matrices the formula for the determinant gets increasingly lengthy, and decreasingly useful for calculation. Once more MATLAB comes to our rescue, since it has a built-in

137

procedure `det` for calculating the determinants of matrices of arbitrary size. In the case of $\mathbf{M}_4$ and $\mathbf{M}_3$, we get

```
>> det(M4)

ans =

    0

>> det(M3)

ans =

    2
```

The following theorem explains exactly what kind of behavior we can expect from a determined system. In particular the results we noticed earlier are predicted by the determinants of the matrices $\mathbf{M}_3$ and $\mathbf{M}_4$.

Theorem. *Let* $\mathbf{A}$ *be an* $n \times n$ *matrix.*

a) *If* $\det \mathbf{A} \neq 0$, *then for every vector* $\mathbf{b}$ *there is a unique vector* $\mathbf{x}$ *such that* $\mathbf{Ax} = \mathbf{b}$. *In particular the only solution to the homogeneous equation* $\mathbf{Ax} = \mathbf{0}$ *is the zero vector* $\mathbf{0}$.

b) *If* $\det \mathbf{A} = 0$, *then there are vectors* $\mathbf{b}$ *such that the system of equations* $\mathbf{Ax} = \mathbf{b}$ *has no solutions. In addition there are nonzero vectors* $\mathbf{x}$ *such that* $\mathbf{Ax} = \mathbf{0}$.

If $\det \mathbf{A} = 0$, we will say that the matrix $\mathbf{A}$ is *singular*. On the other hand, if $\det \mathbf{A} \neq 0$, we will say that $\mathbf{A}$ is *nonsingular*.

If the matrix $\mathbf{A}$ is nonsingular, then we can always solve the equation $\mathbf{Ax} = \mathbf{b}$ uniquely. MATLAB has a convenient way to do this. It is only necessary to enter x=A\b. Trying this, we get

```
>> M3\b

ans =

       1
   1/1125899906842624
      -3
```

the very large denominator in the second term indicates that MATLAB is approximating 0. (Look at this answer in `format long`.) Thus this solution agrees with the one found earlier. On the

138

other hand, if we try

```
>> M4\b
```

Warning: Matrix is singular to working precision.

ans =

```
        1/0
        1/0
        1/0
```

we see that MATLAB recognizes the fact that M_4 is a singular matrix.

The backslash command A\b is deceptively simple. It looks as if we are dividing on the left by the matrix **A**. In a way this is true, but in order to implement that division, MATLAB has to go through a series of computations involving row operations which is very similar to those we discussed earlier in this chapter.

If the matrix **A** is nonsingular, then it has an *inverse*. This is another matrix **B** such that $\mathbf{AB} = \mathbf{BA} = \mathbf{I}$. Here **I** is the *identity matrix*, which has all ones along the diagonal, and zeros off the diagonal. In general it is time-consuming to calculate the inverse of a matrix, but again MATLAB can do this for us. We enter B=inv(A). For example,

```
>> B=inv(M3)
```

B =

```
        -1              2              4
         1           -1/2             -3
        -1              1              3
```

You can verify that $\mathbf{BM_3} = \mathbf{M_3B} = \mathbf{I}$. By the way, in MATLAB the command eye(n) is used to designate the $n \times n$ identity matrix.

The nullspace of a matrix

The *nullspace* of a matrix **A** is the set of all vectors **x** such that $\mathbf{Ax} = \mathbf{0}$. According to the theorem in the previous section, if **A** is nonsingular, then the nullspace of **A** consists of just the zero vector **0**. On the other hand, the nullspace of $\mathbf{M_4}$ consists of all multiples of the vector

$$\begin{pmatrix} 1 \\ -1 \\ 1 \end{pmatrix}.$$

139

Finding the nullspace of a matrix $\mathbf{A}$ involves finding all solutions to the system of homogeneous equations $\mathbf{Ax} = \mathbf{0}$. From this point of view, there is nothing new involved except some new terminology. There is one minor simplification that can be made in the method. Notice that in the case of homogeneous equations, constructing the augmented matrix involves adding a column vector of all zeros to the matrix $\mathbf{A}$. In addition notice that the three row operations will leave this vector of zeros unchanged. Thus augmenting the coefficient matrix is unnecessary, since all of the information is contained in the coefficient matrix itself.

For example consider the matrix

```
>> A=[12 -8 -4 30;3 -2 -1 6;18 -12 -6 42;6 -4 -2 15]

A =

        12        -8        -4        30
         3        -2        -1         6
        18       -12        -6        42
         6        -4        -2        15
```

Applying `rref` we get

```
>> rref(A)

ans =

         1      -2/3      -1/3         0
         0         0         0         1
         0         0         0         0
         0         0         0         0
```

Thus the system $\mathbf{Ax} = \mathbf{0}$ is equivalent to the system

$$x_1 - (2/3)x_2 - (1/3)x_3 = 0$$
$$x_4 = 0.$$

Consequently x_2 and x_3 are free variables. We set $x_2 = t$ and $x_3 = s$, and we solve to find $x_1 = (2/3)t + (1/3)s$ and $x_4 = 0$. Thus the nullspace of $\mathbf{A}$ consists of all vectors of the form

$$x = \begin{pmatrix} (2/3)t + (1/3)s \\ t \\ s \\ 0 \end{pmatrix} = t \begin{pmatrix} 2/3 \\ 1 \\ 0 \\ 0 \end{pmatrix} + s \begin{pmatrix} 1/3 \\ 0 \\ 1 \\ 0 \end{pmatrix}.$$

If we set

$$\mathbf{v}_1 = \begin{pmatrix} 2/3 \\ 1 \\ 0 \\ 0 \end{pmatrix} \quad \text{and} \quad \mathbf{v}_2 = \begin{pmatrix} 1/3 \\ 0 \\ 1 \\ 0 \end{pmatrix},$$

140

then the nullspace consists of all vectors of the form $t\mathbf{v}_1 + s\mathbf{v}_2$. Such a vector is called a *linear combination* of $\mathbf{v}_1$ and $\mathbf{v}_2$. Since the numbers t and s are arbitrary, the nullspace consists of all linear combinations of $\mathbf{v}_1$ and $\mathbf{v}_2$. In particular, the nullspace conatins infinitely many vectors, but if we know $\mathbf{v}_1$ and $\mathbf{v}_2$ we know them all.

The results in this example exemplify what happens in general. The nullspace of a matrix will consist of all linear combinations of a small number of explicit vectors. If these vectors are linearly independent (see the next paragraph), they are called a *basis* of the nullspace, and the nullspace is said to be *spanned* by the basis vectors. If we know a basis for a nullspace, then we know all of the vectors in the nullspace. For this reason we will almost always describe a nullspace by giving a basis.

A finite set of vectors is said to be *linearly independent* if the only linear combination which is equal to the zero vector is the one where all of the coefficients are equal to zero, i.e., the vectors $\mathbf{v}^1, \mathbf{v}^2, \cdots, \mathbf{v}^n$ are linearly independent if $0 = c_1 \mathbf{v}^1 + c_2 \mathbf{v}^2 + \cdots + c_k \mathbf{v}^k$ implies that $c_1 = c_2 = \cdots = c_k = 0$. A set of vectors which is not linearly independent is said to be *linearly dependent*.

Using `rref` it is easy to write down a basis for the nullspace of a matrix. There is one basis vector for each free variable (i.e. free column in the reduced row echelon form). We give that free variable a nonzero value, set each of the other free variables equal to zero, and then solve for the pivot variables. It is often convenient to choose the nonzero value so that the answers come out to be integers. In our example we could first set $x_2 = 3$ and $x_3 = 0$, and get the vector

$$\mathbf{w}_1 = \begin{pmatrix} 2 \\ 3 \\ 0 \\ 0 \end{pmatrix}.$$

Then we set $x_3 = 3$ and $x_2 = 0$, and get

$$\mathbf{w}_2 = \begin{pmatrix} 1 \\ 0 \\ 3 \\ 0 \end{pmatrix}.$$

The vectors $\mathbf{w}_1$ and $\mathbf{w}_2$ form a basis for the nullspace.

Notice that the pair of vectors $\mathbf{v}_1$ and $\mathbf{v}_2$ and the pair $\mathbf{w}_1$ and $\mathbf{w}_2$ are both bases for the same nullspace. Hence we see that there can be more than one basis for a nullspace. In this case the two bases look very similar, but this does not have to be the case. In fact there is not a basis that is preferred to all others, so any one basis is as good as any other.

However, it is true that every basis for a nullspace contains the same number of vectors. This number is by definition the *dimension* of the nullspace. Thus the nullspace in our example has dimension equal to 2.

A basis for a nullspace can also be found using the MATLAB command `null`. Entering `null(A)` results in

```
>> null(A)

ans =

    0.5976    -0.0019
    0.7185     0.4450
    0.3558    -0.8955
    0.0000    -0.0000
```

These two column vectors constitute yet another basis for the nullspace of **A**.

The output of `null` is normalized so that the sum of the squares of the entries of the vectors are equal to 1, and the dot products of different vectors is equal to 0. This normalization is often useful, but the result is that even when we expect to find basis vectors with entries which are small rational numbers, `null` will not produce them. If we want rational output we are required to use `rref`.

The exception to this rule occurs when the dimension of the nullspace is 1. In this case the basis consists of one vector. Dividing that vector by the entry which is smallest in absolute value will result in a vector of rational numbers.

Linear dependence and independence

Consider the vectors

$$\mathbf{v}^1 = \begin{pmatrix} 1 \\ -1 \\ 4 \end{pmatrix} \quad \mathbf{v}^2 = \begin{pmatrix} -2 \\ 4 \\ 5 \end{pmatrix} \quad \text{and} \quad \mathbf{v}^3 = \begin{pmatrix} -2 \\ 10 \\ 44 \end{pmatrix}.$$

How can one tell whether or not these are linearly dependent? If they are linearly dependent, how do we find nonzero coefficients c_1, c_2, and c_3 such that $c_1\mathbf{v}^1 + c_2\mathbf{v}^2 + c_3\mathbf{v}^3 = \mathbf{0}$?

According to the definition, to answer these questions we must look for coefficients c_1, c_2, and c_3 which satisfy the equation $c_1\mathbf{v}^1 + c_2\mathbf{v}^2 + c_3\mathbf{v}^3 = 0$. Let **V** denote the matrix whose column vectors are $\mathbf{v}^1$, $\mathbf{v}^2$, and $\mathbf{v}^3$. We will denote this by $\mathbf{V} = [\mathbf{v}^1, \mathbf{v}^2, \mathbf{v}^3]$. Notice that

$$c_1\mathbf{v}^1 + c_2\mathbf{v}^2 + c_3\mathbf{v}^3 = \mathbf{V}\begin{pmatrix} c_1 \\ c_2 \\ c_3 \end{pmatrix}.$$

Thus if we put the coefficients for which we are searching into a column vector, that vector must belong to the nullspace of the matrix **V**.

The observation of the previous paragraph leads us to the following conclusions. If the nullspace of **V** contains a nonzero vector, then the columns of **V** are linearly dependent, and the elements of the nonzero vector form the coefficients of a non-trivial linear combination of the columns of **V** which is equal to the zero vector. On the other hand, if the nullspace of **V** contains only the zero vector, then the columns of **V** are linearly independent. This completely answers our questions.

Let's see how this works for our example. We will do all of our work in MATLAB.

```
>> v1=[1 ,-1,4]';v2=[-2,4,5]';v3=[-2,10,44]';
>> V=[v1,v2,v3];
>> null(V)

ans =

    0.8242
    0.5494
   -0.1374
```

Since there is a nonzero vector in the nullspace, we conclude that our vectors are linearly dependent. The elements of our answer are the coefficients we need. They can be made into integers by dividing by the entry which is smallest in absolute value.

```
>> c=ans/ans(3)

c =

   -6.0000
   -4.0000
    1.0000
```

It is easily verified that $-6\mathbf{v}^1 - 4\mathbf{v}^2 + \mathbf{v}^3 = 0$.

Let's change the last entry in $\mathbf{v}^3$ to 43, and see if that changes things.

```
>> v3(3)=43;
>> V=[v1,v2,v3];
>> null(V)

ans =

    []
```

The symbol [] means that there are no nonzero vectors in the nullspace, so we conclude that the column vectors

$$\mathbf{v}^1 = \begin{pmatrix} 1 \\ -1 \\ 4 \end{pmatrix} \quad \mathbf{v}^2 = \begin{pmatrix} -2 \\ 4 \\ 5 \end{pmatrix} \quad \text{and} \quad \mathbf{v}^3 = \begin{pmatrix} -2 \\ 10 \\ 43 \end{pmatrix}$$

are linearly independent.

Although we only dealt with a couple of examples in this section, everything is completely general. These techniques will allow you to decide the dependence or independence of any set of vectors.

Exercises

When doing these exercises, it is a very good idea to use the MATLAB `diary` command, which was explained in Chapter 1.

1. Find the general solution to each of the following systems of linear equations using the method of row operations. You may use MATLAB to perform the operations, but in your submission show all of the operations that you perform. In other words, do not use `rref`.

a)
$$\begin{aligned} -5x + 14y &= -47 \\ -7x + 16y &= -55 \end{aligned}$$

b)
$$\begin{aligned} 2x_1 - 5x_2 + 3x_3 &= 8 \\ 4x_1 + 3x_2 - 7x_3 &= -3 \end{aligned}$$

c)
$$\begin{aligned} -6x - 8y + 8z &= -30 \\ 9x + 11y - 8z &= 33 \\ 9x + 9y - 6z &= 27 \end{aligned}$$

d)
$$\begin{aligned} -19x_1 - 128x_2 + 81x_3 + 38x_4 &= 0 \\ 8x_1 + 61x_2 - 27x_3 - 16x_4 &= 0 \\ 2x_1 + 4x_2 + 12x_3 - 4x_4 &= 0 \\ -8x_1 - 16x_2 + 12x_3 + 16x_4 &= 0 \end{aligned}$$

2. Find the general solution to each of the following systems of linear equations. In this problem you may use `rref`. You may also use the backslash operation (`A\b`) if it applies.

a)
$$\begin{aligned} -3x - 9y + 6z &= 15 \\ 2x + 9y + z &= -16 \end{aligned}$$

144

$$8x - 10y - 228z = -112$$

b) $$2x - y - 4z = -16$$

$$4x - 5y - 14z = -56$$

$$-19x_1 - 128x_2 + 81x_3 + 38x_4 = 3$$

$$8x_1 + 61x_2 - 27x_3 - 16x_4 = 5$$

c) $$2x_1 + 4x_2 + 12x_3 - 4x_4 = -21$$

$$-8x_1 - 16x_2 + 12x_3 + 16x_4 = -4$$

d) $\mathbf{Ax} = \mathbf{b}$ where $\mathbf{A} = \begin{pmatrix} 2 & 0 & 2 \\ 1 & 0 & 1 \\ -7 & 12 & 5 \end{pmatrix}$, and $\mathbf{b} = \begin{pmatrix} 6 \\ 3 \\ 27 \end{pmatrix}$

e) $\mathbf{Ax} = \mathbf{b}$ where $\mathbf{A} = \begin{pmatrix} 2/3 & 0 & -2/3 & 1/3 \\ 11/3 & -3 & -20/3 & 1/3 \\ -3 & 3 & 6 & 0 \\ -16/3 & 6 & 34/3 & 1/3 \end{pmatrix}$, and $\mathbf{b} = \begin{pmatrix} 1/3 \\ -80/3 \\ 27 \\ 163/3 \end{pmatrix}$

f) $\mathbf{Ax} = \mathbf{b}$ where $\mathbf{A} = \begin{pmatrix} -23 & 26 & -42 & -32 & -90 \\ -2 & 1 & 0 & -3 & -4 \\ -17 & 19 & -28 & -22 & -63 \\ -14 & 14 & -24 & -16 & -52 \\ 18 & -20 & 32 & 23 & 69 \end{pmatrix}$ and $\mathbf{b} = \begin{pmatrix} -6 \\ -2 \\ -3 \\ -2 \\ 3 \end{pmatrix}$

3. For each of the following matrices find a basis for the nullspace. What is the dimension in each case?

a) $\begin{pmatrix} 2 & 2 \\ -1 & -1 \end{pmatrix}$ b) $\begin{pmatrix} 0 & 4 & -2 \\ 1 & -2 & 1 \\ 3 & -10 & 5 \end{pmatrix}$ c) $\begin{pmatrix} 2 & -4 & -9 \\ 0 & 0 & -2 \\ 0 & 0 & 1 \end{pmatrix}$

d) $\begin{pmatrix} 12 & -5 & -14 & -9 \\ 18 & -7 & -22 & -12 \\ 12 & -6 & -12 & -9 \\ -16 & 8 & 16 & 11 \end{pmatrix}$ e) $\begin{pmatrix} -3 & 2 & 5 & 2 \\ 6 & -2 & -8 & -2 \\ -4 & 2 & 6 & 2 \\ -4 & 2 & 6 & 2 \end{pmatrix}$

f) $\begin{pmatrix} 6 & -4 & 2 & -12 & -8 \\ 6 & -4 & 2 & -12 & -8 \\ 29 & -14 & 11 & -54 & -36 \\ 13 & -6 & 5 & -24 & -16 \\ -10 & 4 & -4 & 18 & 12 \end{pmatrix}$

4. For each of the following set of vectors, determine if they are linearly dependent or independent. If they are dependent, find a nonzero linear combination which is equal to the zero vector.

a) $\begin{pmatrix} 1 \\ 2 \end{pmatrix}$ and $\begin{pmatrix} -3 \\ -6 \end{pmatrix}$

b) $\begin{pmatrix} 1 \\ 2 \end{pmatrix}$ and $\begin{pmatrix} -3 \\ -5 \end{pmatrix}$

c) $\begin{pmatrix} 1 \\ 2 \\ 0 \end{pmatrix}$ and $\begin{pmatrix} -3 \\ -6 \\ 1 \end{pmatrix}$

d) $\begin{pmatrix} 1 \\ 1 \\ 1 \end{pmatrix}$, $\begin{pmatrix} 1 \\ -1 \\ 1 \end{pmatrix}$, and $\begin{pmatrix} 5 \\ 0 \\ 5 \end{pmatrix}$

e) $\begin{pmatrix} 1 \\ 1 \\ 1 \end{pmatrix}$, $\begin{pmatrix} 1 \\ -1 \\ 1 \end{pmatrix}$, and $\begin{pmatrix} 5 \\ 1 \\ 5 \end{pmatrix}$

f) $\begin{pmatrix} 1 \\ 0 \\ 1 \\ 0 \end{pmatrix}$, $\begin{pmatrix} 0 \\ 1 \\ 1 \\ 1 \end{pmatrix}$, and $\begin{pmatrix} 5 \\ -6 \\ -1 \\ -6 \end{pmatrix}$

g) $\begin{pmatrix} 1 \\ 0 \\ 1 \\ 0 \end{pmatrix}$, $\begin{pmatrix} 0 \\ 1 \\ 1 \\ 1 \end{pmatrix}$, and $\begin{pmatrix} 5 \\ -6 \\ 0 \\ -6 \end{pmatrix}$

9. Homogeneous Linear Systems of ODEs

The use of Matlab can significantly reduce the amount of work involved in finding solutions to homogeneous linear systems of ordinary differential equations with constant coefficients, i.e. equations of the form

$$\mathbf{x}' = \mathbf{A}\mathbf{x} \tag{1}$$

where $\mathbf{A}$ is an $n \times n$ matrix and $\mathbf{x}(t)$ is an n-vector of functions. In this chapter we will show how to use the algebra developed in the previous chapter to solve these systems.

Homogeneous systems and eigenvalues

If there is one consistent theme that comes out of the study of ordinary differential equations, it is that when we have a linear homogeneous equation with constant coefficients, we should look for exponential solutions, i.e., solutions of the form $e^{rt}c$, where c is a constant. In our case, where we want to find solutions for the system in (1), our solution must be a vector valued function, and, therefore, the constant must be a vector of constants. Accordingly we are led to look for solutions of the form

$$\mathbf{x}(t) = e^{rt}\mathbf{v}, \tag{2}$$

where $\mathbf{v}$ is an ordinary vector.

Let's see what is required for the function in (2) to be a solution to (1). First of all

$$\mathbf{x}'(t) = re^{rt}\mathbf{v} = e^{rt}r\mathbf{v}. \tag{3}$$

On the other hand,

$$\mathbf{A}\mathbf{x}(t) = e^{rt}\mathbf{A}\mathbf{v}. \tag{4}$$

In order for (3) and (4) to be equal, we must have

$$\mathbf{A}\mathbf{v} = r\mathbf{v}. \tag{5}$$

Equation (5) expresses constraints on the number r and the vector $\mathbf{v}$ in order that (2) be a solution to (1). This leads to the following definition. Remember that we want a nonzero solution.

Definition. *Let $\mathbf{A}$ be an $n \times n$ matrix. A number r is said to be an eigenvalue for $\mathbf{A}$ if there is a nonzero vector $\mathbf{v}$ such that $\mathbf{A}\mathbf{v} = r\mathbf{v}$. If r is an eigenvalue for $\mathbf{A}$, then any vector $\mathbf{v}$ which satisfies $\mathbf{A}\mathbf{v} = r\mathbf{v}$ is called an eigenvector for $\mathbf{A}$ associated to the eigenvalue r.*

It is already apparent that eigenvalues and eigenvectors are important in the study of first order systems of differential equations. For that reason alone their study is worthwhile, but in fact applications of these ideas occur in many areas.

Next let's turn to analyzing the definition. We will use $\mathbf{I}$, the identity matrix of the same size as $\mathbf{A}$. This matrix has ones along the diagonal and zeros elsewhere. It has the property that $\mathbf{I}\mathbf{v} = \mathbf{v}$ for all vectors $\mathbf{v}$. Using (5) we see that if r is an eigenvalue of $\mathbf{A}$, and $\mathbf{v}$ is an associated eigenvector, then

$$
\begin{aligned}
0 &= \mathbf{A}\mathbf{v} - r\mathbf{v} \\
&= \mathbf{A}\mathbf{v} - r\mathbf{I}\mathbf{v} \\
&= (\mathbf{A} - r\mathbf{I})\,\mathbf{v}
\end{aligned}
$$

Thus, $\mathbf{v}$ must be in the nullspace of the matrix $\mathbf{A} - r\mathbf{I}$. From the theorem in the previous chapter we know that there is a nonzero vector in the nullspace of $\mathbf{A} - r\mathbf{I}$ if and only if $\det(\mathbf{A} - r\mathbf{I}) = 0$. This is a necessary and sufficient condition for r to be an eigenvalue of $\mathbf{A}$. Notice that this condition involves only the eigenvalue r; it does not require us to find an eigenvector at the same time.

The function

$$
p(r) = \det(\mathbf{A} - r\mathbf{I}) \tag{6}
$$

is a polynomial of degree n if $\mathbf{A}$ is an $n \times n$ matrix. It is called the *characteristic polynomial* of $\mathbf{A}$. We see that the eigenvalues of $\mathbf{A}$ are precisely the roots of the characteristic polynomial.

Let's look at an example. Let

$$
\mathbf{A} = \begin{pmatrix} -3 & 1 & -3 \\ -8 & 3 & -6 \\ 2 & -1 & 2 \end{pmatrix}. \tag{7}
$$

Then

$$
\mathbf{A} - r\mathbf{I} = \begin{pmatrix} -3 - r & 1 & -3 \\ -8 & 3 - r & -6 \\ 2 & -1 & 2 - r \end{pmatrix}.
$$

If you are good at calculating determinants, you will find that the characteristic polynomial is $p(r) = r^3 - 2r^2 - r + 2$.

However, there is a lot of computation required to calculate that determinant, so why not let MATLAB do it? The command `poly(A)` will display a vector containing the coefficients of the characteristic polynomial.

```
>> A=[-3 1 -3;-8 3 -6;2 -1 2];
>> p = poly(A)

p =

    1        -2        -1         2
```

148

To find the eigenvalues we must find the roots of the characteristic polynomial, which is equivalent to factoring the polynomial. If you are a good factorer, you will find that

$$p(r) = r^3 - 2r^2 - r + 2 = (r - 1)(r + 1)(r - 2).$$

Thus the eigenvalues of **A** are 1, -1, and 2.

Once again MATLAB can save us time. If we have calculated the coefficients of the characteristic polynomial as the vector p, then we can use `roots(p)` to compute the roots.

```
>> roots(p)

ans =


     2
     1
    -1
```

There are several commands in the Symbolic Toolbox which are useful in studying these issues. For example, the command `charpoly(A)` will display the characteristic polynomial of the matrix **A**.

```
>> ps = charpoly(A)

ps =

x^3-2*x^2-x+2
```

Then the eigenvalues can be found by using the `solve` command.

```
>> solve(ps)

ans =

[ 1]
[ 2]
[-1]
```

The Symbolic Toolbox has another command, `factor`, which is also of interest here. This command will factor the characteristic polynomial.

```
>> factor(ps)

ans =

(x-1)*(x-2)*(x+1)
```

Eigenvalues and eigenvectors using MATLAB

Having found the eigenvalues, we must now find the associated eigenvectors. If r is an eigenvalue for $\mathbf{A}$, then the eigenvectors are the vectors which are in the null space of $\mathbf{A} - r\mathbf{I}$. We discussed how to find the nullspace of a matrix in the previous chapter, so presumably we know how to do this as well.

The complete process, as we have described it, has three steps. Find the characteristic polynomial; find the eigenvalues, which are the roots of the characteristic polynomial; and then, for each eigenvalue r, find the associated eigenvectors by finding the nullspace of $\mathbf{A} - r\mathbf{I}$. The third step can be accomplished either by row operations on the matrix $\mathbf{A} - r\mathbf{I}$, or by using the command `null` applied to $\mathbf{A} - r\mathbf{I}$.

The three step process can be replaced by one using the MATLAB command `eig`. First of all, the command `eig(A)` will display the eigenvalues of $\mathbf{A}$.

```
>> eig(A)

ans =

    2
   -1
    1
```

However, the command `[V,E] = eig(A)` will output two matrices. E will be a diagonal matrix with the eigenvalues along the diagonal, and V will have the associated eigenvectors as its column vectors. For example, using `format short` we get the following:

```
>> [V,E]=eig(A)

V =

    0.4082    0.4472   -0.5774
    0.8165    0.8944   -0.5774
   -0.4082    0.0000    0.5774


E =

    2.0000         0         0
         0   -1.0000         0
         0         0    1.0000
```

The first column of V is an eigenvector associated with the first element of the diagonal of E, and so on.

Eigenvectors are not unique. Any nonzero multiple of an eigenvector will also be an eigenvector associated to the same eigenvalue. Thus eigenvectors can be normalized in a variety of ways. The eigenvectors that are produced by eig are normalized so that the sum of the squares of the elements is equal to 1. For many purposes this is a natural normalization. However, if the matrix has integer entries, and if the eigenvalues are also integers, we would expect to be able to find eigenvectors with integer entries. Sometimes we can get such eigenvectors by easy manipulation of the column vectors in V. For example, we often get what we want by dividing a column vector by its smallest entry. With the first column of the matrix V we get:

```
>> format rat
>> v1=V(:,1)/V(1,1)

v1 =

        1
        2
       -1
```

We can check that this is an eigenvector:

```
>> A*v1

ans =

        2
        4
       -2
```

This is clearly 2*v1.

Although this works for all of the columns in V, it does not work for every matrix. If we want to find eigenvectors with entries which are integers or small rational numbers, we may need to find the eigenvalues with eig, and then compute the nullspace as we did in the previous chapter.

If you have the Symbolic Toolbox available, there is yet another approach to the problem of finding eigenvalues and eigenvectors. This method works especially well for relatively small matrices with integer or rational entries. The command eigensys(A) yields a symbolic vector containing the eigenvalues of A.

```
>> eigensys(A)
```

```
ans =

[ 1]
[ 2]
[-1]
```

The command `[V,E] = eigensys(A)` yields a symbolic matrix V containing the eigenvectors as columns, and a symbolic vector E containing the eigenvectors.

```
>> [V,E] = eigensys(A)

V =

[-1, -1, 1]
[-1, -2, 2]
[ 1,  1, 0]

E =

[ 1]
[ 2]
[-1]
```

These symbolic matrices can be converted to numeric matrices by the command `numeric`. For example,

```
>> V=numeric(V)

V =

        -1          -1           1
        -1          -2           2
         1           1           0
```

The vector V is now a standard numerical vector and can be used for other computations.

It is important to understand the differences in the way that `eig` and `eigensys` work. The function `eig` is a *numerical* routine. It uses floating point arithmetic to do its computation. At every step it is rounding off and making approximations. The result is always a highly accurate, but still approximate, answer. However, it will work on any matrix you give it. A nice test is to generate a 100×100 matrix and see how long it takes MATLAB to calculate the eigenvalues and eigenvectors. The command `A=rand(100);` will generate such a matrix.

On the other hand, `eigensys` is a *symbolic* routine. It computes its answers using highly sophisticated algebra, involving finding the roots of polynomials. As a result, you can count on the solution being 100% accurate. However, `eigensys` will not always give the most usable results. Let's look at some examples.

For the matrix

$$A = \begin{pmatrix} -9 & 4 \\ -2 & 2 \end{pmatrix} \tag{8}$$

`eigensys` gives us

```
>> [V,E] = eigensys(A)

V =

[                1,                1]
[1/4*E(1)+9/4, 1/4*E(2)+9/4]

E =

[-7/2+1/2*89^(1/2)]
[-7/2-1/2*89^(1/2)]
```

Thus the eigenvalues are $(-7\pm\sqrt{89})/2$. These are the exact values, involving no approximation. The entries of the eigenvectors are also complicated — so much so that it is desirable to express the answer in terms of the eigenvalues, i.e. the components of the vector E.

On the other hand, `eig` yields

```
>> [V1,E1]=eig(A)

V1 =

    -0.9814    -0.3646
    -0.1921    -0.9312


E1 =

    -8.2170         0
          0    1.2170
```

This time we get decimal approximations for the eigenvalues and the eigenvectors. The two answers are very close:

```
>> format long
>> E1

E1 =

   -8.21699056602830                       0
                   0    1.21699056602830


>> numeric(E)

ans =

    1.21699056602830
   -8.21699056602830
```

We see that, up to the 14 decimal places permitted by this format, the two answers agree.

However, there is an additional difficulty with `eigensys`. Because it depends on algebraic methods, and because there are no general algebraic methods for finding the roots of polynomials of degree five or larger, `eigensys` will not be able to compute symbolically the eigenvectors of a matrix larger than 4×4, unless the characteristic polynomial can be factored into polynomials of degree less than or equal to four with coefficients which are rational numbers. This is not a usual occurrence.

For cases when the symbolic computation fails, the Symbolic Toolbox provides an alternative. The command `[V,E]=eigensys(vpa(A))` will result in numerical values for the eigenvalues and eigenvectors.

It is time to sum up our progress. Our goal is to find the general solution to equations like (1). The general theory tells us that if **A** is an $n \times n$ matrix, then there are n independent solutions, and the general solution is a linear combination of these. So far we have discovered how to find specific solutions of the type given by (2). We can find one such for every eigenvalue of **A**. Since the eigenvalues are the roots of the characteristic polynomial, which is of degree n, in general there will be n roots, and therefore n solutions to (1). Thus it seems we can find the needed number of solutions.

We will find that this analysis is correct, except in some cases when the characteristic polynomial has multiple roots. In such cases, we will still get at least one solution for each distinct eigenvalue, but this will not be enough. We will have to work much harder to find the additional solutions.

In the remainder of the chapter we will compute specific examples. In addition we will address the issue of solving initial value problems. When the eigenvalues are complex, the

154

solutions of the form (2) will be complex valued, so we will have to do some algebra to find real valued solutions. Finally, when the eigenvalues have multiplicity greater than one there will be extra work.

We end this section with a mathematical aside. We mentioned earlier the three step process for calculating eigenvalues and eigenvectors. In fact, the function eig does not proceed by this process. It uses an algorithm, called the *QR algorithm*, which computes the eigenvalues and the eigenvectors all at once. The QR algorithm is a triumph of modern computational linear algebra. It works so well that MATLAB reverses the three step process. When you ask MATLAB to compute the roots of a polynomial, it first finds a matrix (called the *companion matrix*) which has the given polynomial as its characteristic polynomial. Then it uses the QR algorithm to compute the eigenvalues of that matrix, which are of course the roots of the polynomial. The interested reader will find the QR algorithm described in modern books on numerical linear algebra.

Real eigenvalues

Before we start something new, let's finish with the examples in the previous section. First we want to find the solution to the system in (1) for the matrix **A** as given in (7).

We have calculated the eigenvalues and eigenvectors of **A** using eig, and these are stored in the matrices E and V respectively. To make things simple, let's first alter V so that the eigenvectors have integer entries. This can be achieved by dividing each column vector by the nonzero entry in that column which is smallest in absolute value.

```
>> V(:,1) = V(:,1)/V(1,1);
>> V(:,2)=V(:,2)/V(1,2);
>> V(:,3)=V(:,3)/V(1,3)
```

```
V =

    1.0000    1.0000    1.0000
    2.0000    2.0000    1.0000
   -1.0000    0.0000   -1.0000
```

Now we are in a position to write down the solutions corresponding to the three eigenvectors. Each is of the form in (2). i.e., $\mathbf{x}^i(t) = e^{r_i t}\mathbf{v}^i$, where r_i is an eigenvalue, and $\mathbf{v}^i$ is an associated eigenvector. Since the eigenvectors are the column vectors in V, we can just read off the solutions. They are

$$\mathbf{x}^1(t) = e^{2t}\mathbf{v}^1 = e^{2t}\begin{pmatrix} 1 \\ 2 \\ -1 \end{pmatrix}$$

155

$$\mathbf{x}^2(t) = e^{-t}\mathbf{v}^2 = e^{-t}\begin{pmatrix} 1 \\ 2 \\ 0 \end{pmatrix}$$

$$\mathbf{x}^3(t) = e^t\mathbf{v}^3 = e^t\begin{pmatrix} 1 \\ 1 \\ -1 \end{pmatrix}.$$

Thus the general solution is

$$\mathbf{x}(t) = c_1\mathbf{x}^1(t) + c_2\mathbf{x}^2(t) + c_3\mathbf{x}^3(t)$$
$$= c_1 e^{2t}\mathbf{v}^1 + c_2 e^{-t}\mathbf{v}^2 + c_3 e^t\mathbf{v}^3 .$$

(9)

Suppose we want to find the specific solution which has the initial value

$$\mathbf{x}(0) = \begin{pmatrix} 2 \\ 2 \\ -3 \end{pmatrix}.$$

From (9) we see that $\mathbf{x}(0) = c_1\mathbf{v}^1 + c_2\mathbf{v}^2 + c_3\mathbf{v}^3$. Notice that with $\mathbf{V} = [\mathbf{v}^1, \mathbf{v}^2, \mathbf{v}^3]$, the matrix whose column vectors are the eigenvectors, we have

$$\mathbf{Vc} = \mathbf{V}\begin{pmatrix} c_1 \\ c_2 \\ c_3 \end{pmatrix} = c_1\mathbf{v}^1 + c_2\mathbf{v}^2 + c_3\mathbf{v}^3.$$

Here $\mathbf{c}$ is used to indicate the vector of constants. We have shown that if $\mathbf{x}$ is given by (9), then $\mathbf{x}(0) = \mathbf{Vc}$. Thus to solve the initial value problem, we must find the vector of constants which satisfies

$$\mathbf{Vc} = \begin{pmatrix} 2 \\ 2 \\ -3 \end{pmatrix}.$$

Nothing could be easier for MATLAB.

```
>> c=V\[2 2 -3]'

c =

    1.0000
   -1.0000
    2.0000
```

So the solution to the initial value problem is

$$\begin{aligned}
\mathbf{x}(t) &= \mathbf{x}^1(t) - \mathbf{x}^2(t) + 2\mathbf{x}^3(t) \\
&= e^{2t}\mathbf{v}^1 - e^{-t}\mathbf{v}^2 + 2e^t\mathbf{v}^3 \\
&= \begin{pmatrix} e^{2t} - e^{-t} + 2e^t \\ 2e^{2t} - 2e^{-t} + 2e^t \\ -e^{2t} - 2e^t \end{pmatrix}.
\end{aligned} \tag{10}$$

Next let's finish the example in (8). We have

```
>> [V,E] = eig(A)

V =

   -0.9814   -0.3646
   -0.1921   -0.9312


E =

   -8.2170         0
         0    1.2170
```

From the discussion in the previous section we know that we can give more precise answers if we are willing to use expressions involving square roots. Instead let's stay with what `eig` gives us, but notice that all of the entries in V are negative. Since the columns of V are the eigenvectors, we will still have eigenvectors if we replace V by -V.

```
>> V=-V

V =

   0.9814   0.3646
   0.1921   0.9312
```

Then we can immediately write down two solutions:

$$\mathbf{x}^1(t) = e^{-8.2170\,t}\begin{pmatrix} 0.9814 \\ 0.1921 \end{pmatrix} \quad \text{and} \quad \mathbf{x}^2(t) = e^{1.2170\,t}\begin{pmatrix} 0.3646 \\ 0.9312 \end{pmatrix}$$

To solve the initial value problem with $\mathbf{x}(0) = \begin{pmatrix} 1 \\ -1 \end{pmatrix}$, we proceed as in the previous example. We know the solution has the form $\mathbf{x}(t) = a_1\mathbf{x}^1(t) + a_2\mathbf{x}^2(t)$, so

$$\begin{pmatrix} 1 \\ -1 \end{pmatrix} = \mathbf{x}(0)a_1\mathbf{x}^1(0) + a_2\mathbf{x}^2(0) = a_1V^1 - a_2V^2 = \mathbf{V}\begin{pmatrix} a_1 \\ a_2 \end{pmatrix}.$$

157

Thus,

```
>> a=V\[1 -1]'

a =

    1.5356
   -1.3907
```

and our solution is

$$\mathbf{x}(t) = 1.5356\,e^{-8.2170\,t}\begin{pmatrix} 0.9814 \\ 0.1921 \end{pmatrix} - 1.3907\,e^{1.2170\,t}\begin{pmatrix} 0.3646 \\ 0.9312 \end{pmatrix}.$$

Let's look at another example which reveals some of the vagaries of using MATLAB. Our matrix will be

```
A =

    15        -6       -18        -6
    -4         5         8         4
    12        -6       -15        -6
     4        -2        -8        -1
```

If we use [V,E]=eig(A), we get a complicated business that fills the screen. We will not attempt to duplicate it here. It would be difficult to get any information from what is there. We can look at the eigenvalues by looking only at the diagonal entries of the matrix E using:

```
>> diag(E)

ans =

   -3
    1
    3    +    1/1616502429429809i
    3    -    1/1616502429429809i
```

This is better. It indicates that MATLAB thinks that the eigenvalues are -3, 1, and two complex numbers with real part equal to 3 and a very small imaginary part. In fact, that imaginary part is so small that we are led to think that it is 0, and that 3 is an eigenvalue with multiplicity two. We can check this conjecture by computing the nullspace of $\mathbf{A} - 3\mathbf{I}$. The

MATLAB notation for the 4×4 identity matrix is `eye(4)`. We proceed as follows:

```
>> I=eye(4);
>> rref(A-3*I)

ans =

    1        -1/2       0          1
    0         0         1          1
    0         0         0          0
    0         0         0          0
```

Since there are two free variables, we see that we can compute two independent vectors in the nullspace, i.e. two eigenvectors corresponding to the eigenvalue 3. This confirms our conjecture. To get eigenvectors $\mathbf{v}$ with integer entries we first choose $v_2 = 2$ and $v_4 = 0$. Then we choose $v_2 = 0$ and $v_4 = 1$. This yields the eigenvectors

$$\begin{pmatrix} 1 \\ 2 \\ 0 \\ 0 \end{pmatrix} \quad \text{and} \quad \begin{pmatrix} -1 \\ 0 \\ -1 \\ 1 \end{pmatrix}$$

associated with the eigenvalue 3.

We could have also used `null(A-3*I)` to compute two eigenvectors associated with the eigenvalue 3, but it is very difficult to get vectors with integer entries that way.

The eigenvectors associated with the other eigenvalues, 1 and -3, can be computed using the matrix **V** as we did in the previous example. They can also be found starting with the commands `null(A-I)` and `null(A-(-3)*I)`, respectively.

The analysis of this case is greatly simplified using the command `eigensys` which is only available with the Symbolic Toolbox.

```
>> [V,E] = eigensys(A)

V =

[ 1,  0, 1,  0]
[-1, -1, 2,  2]
[ 1,  0, 0, -1]
[ 1,  1, 0,  1]
```

E =

[-3]
[1]
[3]
[3]

This gives the total answer about the eigenvalues and eigenvectors. From this we can read off the solutions corresponding to the eigenvectors.

$$\mathbf{x}^1(t) = e^{-3t}\begin{pmatrix}1\\-1\\1\\1\end{pmatrix}, \quad \mathbf{x}^2(t) = e^t\begin{pmatrix}0\\-1\\0\\1\end{pmatrix}, \quad \mathbf{x}^3(t) = e^{3t}\begin{pmatrix}1\\2\\0\\0\end{pmatrix} \quad \text{and} \quad \mathbf{x}^4(t) = e^{3t}\begin{pmatrix}0\\2\\-1\\1\end{pmatrix}.$$

Notice, however, that the eigenvectors associated with the eigenvalue 3 are not the same as those found earlier. This is typical for eigenvalues of multiplicity greater than 1. What is found by any method in such a case is a basis for the eigenvalues, and bases are not unique. Both answers are correct.

Complex eigenvalues

Consider the matrix

$$A = \begin{pmatrix}-1 & 0 & 2\\2 & 3 & -6\\-2 & 0 & -1\end{pmatrix}.$$

We will compute the general solution to the differential system $\mathbf{x}' = \mathbf{Ax}$.

Using [V,E]=eig(A) we have the eigenvalues and eigenvectors of A displayed on the screen of the Command Window. The actual display is too large to include here. Instead we will display it in pieces. First to find the eigenvalues, we use E = eig(A), and

```
>> diag(E)

ans =

        3
       -1        +        2i
       -1        -        2i
```

Thus the eigenvalues are 3, and the complex conjugates $-1 \pm 2i$.

The first column of V will be an eigenvector associated with eigenvalue 3.

```
>> v1 = V(:,1)

v1 =

     0
     1
     0
```

The associated solution is

$$\mathbf{x}^1(t) = e^{3t} \begin{pmatrix} 0 \\ 1 \\ 0 \end{pmatrix} = \begin{pmatrix} 0 \\ e^{3t} \\ 0 \end{pmatrix}.$$

Notice that

$$\mathbf{x}^1(0) = \begin{pmatrix} 0 \\ 1 \\ 0 \end{pmatrix},$$

which is the vector $v1$.

We can deal with the second and third eigenvalues, which are complex conjugates, at one time. Notice that the eigenvectors are also complex conjugates. We will concentrate on the second eigenvalue $-1 + 2i$, and its eigenvector $\mathbf{v}^2$, which we choose so that it has integers as entries.

```
>> v2 = 2*V(:,2)

v2 =

     1
    -1        +      1i
     0        +      1i
```

Corresponding to this eigenpair we have a solution to the differential system.

$$\mathbf{z}(t) = e^{(-1+2i)t} \begin{pmatrix} 1 \\ -1 + i \\ i \end{pmatrix}$$

Expanding this out using Euler's formula, we get

$$\mathbf{z}(t) = e^{-t} \begin{pmatrix} \sin 2t - i\cos 2t \\ (\cos 2t - \sin 2t) + i(\cos 2t + \sin 2t) \\ \cos 2t + i\sin 2t \end{pmatrix}.$$

161

The theory assures us that the real and imaginary parts of this function will be solutions to the original equation. Hence we get two solutions

$$\mathbf{x}^2(t) = e^{-t}\begin{pmatrix} \sin 2t \\ \cos 2t - \sin 2t \\ \cos 2t \end{pmatrix}, \qquad \mathbf{x}^3(t) = e^{-t}\begin{pmatrix} -\cos 2t \\ \cos 2t + \sin 2t \\ \sin 2t \end{pmatrix}.$$

Notice that

$$\mathbf{x}^2(0) = \begin{pmatrix} 0 \\ 1 \\ 1 \end{pmatrix} \quad \text{and} \quad \mathbf{x}^3(0) = \begin{pmatrix} -1 \\ 1 \\ 0 \end{pmatrix}.$$

The general solution is

$$\mathbf{x}(t) = a_1\mathbf{x}^1(t) + a_2\mathbf{x}^2(t) + a_3\mathbf{x}^3(t).$$

Finally we look for the solution to the initial value problem with $\mathbf{x}(0) = \mathbf{c}$, where

```
>> c=[4 -5 9]'

c =


    4
   -5
    9
```

The solution is of the form $\mathbf{x}(t) = a_1\mathbf{x}^1(t) + a_2\mathbf{x}^2(t) + a_3\mathbf{x}^3(t)$, so we must have $\mathbf{c} = a_1\mathbf{x}^1(0) + a_2\mathbf{x}^2(0) + a_3\mathbf{x}^3(0)$. This means that we must have $\mathbf{Wa} = \mathbf{c}$, where the a's are the components of the vector $\mathbf{a} = (a_1, a_2, a_3)$, and where $\mathbf{W} = [\mathbf{x}^1(0), \mathbf{x}^2(0), \mathbf{x}^3(0)]$ is the matrix whose column vectors are the values of the solutions at $t = 0$. These are easily evaluated to give

```
>> W=[0 1 0;1 -1 1;0 0 1]

W =


    0        1        0
    1       -1        1
    0        0        1
```

and then

```
>> a=W\c

a =


  -10
    4
    9
```

162

Hence the solution is $\mathbf{x}(t) = -10\mathbf{x}^1(t) + 4\mathbf{x}^2(t) + 9\mathbf{x}^3(t)$.

Warning: In the previous chapter we learned to use `rref` to compute the basis of a nullspace, and, therefore, to compute eigenvectors. Although this method works fine for real matrices, `rref` has a bug which sometimes will result in errors for complex matrices. It is best not to use `rref` with complex matrices.

Eigenvalues with multiplicity larger than 1

Consider the matrix

$$A = \begin{pmatrix} 13 & 11 \\ -11 & -9 \end{pmatrix}.$$

If we use MATLAB to compute the eigenvalues,

```
>> eig(A)

ans =

    2
    2
```

we find that 2 is an eigenvalue of multiplicity 2. This is confirmed if we use MATLAB to compute the characteristic polynomial:

```
>> poly(A)

ans =

    1    -4     4
```

We get

$$r^2 - 4r + 4 = (r-2)^2.$$

On the other hand, when we look for the eigenvectors associated with the eigenvalue 2,

```
>> rref(A-2*eye(2))

ans =

    1    1
    0    0
```

163

we see that, up to multiplication by a constant, $\mathbf{v}_1 = \begin{pmatrix} 1 \\ -1 \end{pmatrix}$ is the only eigenvector.

We need some terminology to describe this phenomenon. If r_j is an eigenvalue of the matrix $\mathbf{A}$, we define the *algebraic multiplicity* of r_j to be the largest power of $(r - r_j)$ which divides the characteristic polynomial of $\mathbf{A}$. We define the *geometric multiplicity* of r_j to be the dimension of the nullspace of $\mathbf{A} - r_j \mathbf{I}$, i.e. the dimension of the eigenspace associated to r_j.

In the case at hand, the algebraic multiplicity of the eigenvalue 2 is 2, and the geometric multiplicity is 1. It can be shown that the geometric multiplicity is always smaller than or equal to the algebraic multiplicity. Here is a case where it is strictly smaller.

When we are solving systems of ode's, we want to find a number of linearly independent solutions associated with an eigenvalue of the matrix that is equal to the *algebraic* multiplicity. Since each eigenvector gives rise to a solution, we know how to find a number of linearly independent solutions associated with an eigenvalue of the matrix that is equal to the *geometric* multiplicity. How do we find the others when the geometric multiplicity is smaller?

We need two facts. First, given any vector $\mathbf{w}$, the solution to the equation $\mathbf{x}' = \mathbf{A}\mathbf{x}$ with initial value $\mathbf{x}(0) = \mathbf{w}$ is given by

$$\mathbf{x}(t) = e^{\mathbf{A}t}\mathbf{w},$$

where

$$e^{\mathbf{A}t} = \sum_{k=0}^{\infty} \frac{t^k}{k!}\mathbf{A}^k.$$

The infinite sum can be shown to converge for all choices of $\mathbf{A}$ and for all t, although the sum is not easy to calculate.

For any number r we have $\mathbf{A}t = rt\mathbf{I} + t(\mathbf{A} - r\mathbf{I})$. Substituting this into the above expression for $\mathbf{x}(t)$, we get

$$\begin{aligned}
\mathbf{x}(t) &= e^{rt} e^{t(\mathbf{A}-r\mathbf{I})}\mathbf{w} \\
&= e^{rt} \left(I + t(\mathbf{A} - r\mathbf{I}) + \frac{1}{2}t^2(\mathbf{A} - r\mathbf{I})^2 + \ldots \right)\mathbf{w} \\
&= e^{rt} \left(\mathbf{w} + t(\mathbf{A} - r\mathbf{I})\mathbf{w} + \frac{1}{2}t^2(\mathbf{A} - r\mathbf{I})^2\mathbf{w} + \ldots \right).
\end{aligned} \tag{11}$$

The infinite series in (11) is difficult to deal with in general. This brings us to the second important fact. If $r = r_j$ is an eigenvalue of $\mathbf{A}$, then there is an integer d, which is no larger than the algebraic multiplicity of r_j, such that the dimension of the nullspace of $(\mathbf{A} - r_j\mathbf{I})^d$ is equal to the algebraic multiplicity of r_j. For each vector $\mathbf{w}$ in this nullspace, we have $(\mathbf{A} - r_j\mathbf{I})^k\mathbf{w} = 0$ whenever $k \geq d$. Thus the infinite series in the last equation in (11) becomes finite, and theoretically at least, we are in business.

164

Let's look at what all of this means in the case at hand. We see that

```
>> (A-2*eye(2))^2

ans =

    0    0
    0    0
```

Thus $(\mathbf{A} - 2\mathbf{I})^2$ is the matrix of all zeros. Consequently the nullspace consists of all vectors, and therefore has dimension equal to 2, as promised. A basis for the nullspace consists of the eigenvector $\mathbf{v}^1 = \begin{pmatrix} 1 \\ -1 \end{pmatrix}$, and any additional vector which is not a multiple of $\mathbf{v}^1$, say $\mathbf{v}^2 = \begin{pmatrix} 1 \\ 0 \end{pmatrix}$. We know that $(\mathbf{A} - 2\mathbf{I})\mathbf{v}^1 = 0$, and we calculate that

$$(\mathbf{A} - 2\mathbf{I})\mathbf{v}^2 = \begin{pmatrix} 11 \\ -11 \end{pmatrix} = 11\mathbf{v}^1.$$

Thus, using (11), we have the following solutions to the system:

$$\mathbf{x}^1(t) = e^{2t\mathbf{A}}\mathbf{v}^1 = e^{2t}\mathbf{v}^1 = \begin{pmatrix} e^{2t} \\ -e^{2t} \end{pmatrix},$$

and

$$\mathbf{x}^2(t) = e^{2t\mathbf{A}}\mathbf{v}^2$$
$$= e^{2t}\left(\mathbf{v}^2 + t(\mathbf{A} - 2\mathbf{I})\mathbf{v}^2\right)$$
$$= e^{2t}\left(\mathbf{v}^2 + 11\,t\mathbf{v}^1\right)$$
$$= \begin{pmatrix} e^{2t}(1 + 11\,t) \\ -11\,te^{2t} \end{pmatrix}.$$

In both cases the series truncates at the spot indicated, since $(\mathbf{A}-2\mathbf{I})\mathbf{v}^1 = 0$, and $(\mathbf{A}-2\mathbf{I})^2\mathbf{v}^2 = 0$.

Let's do a more complicated example. Let

$$\mathbf{A} = \begin{pmatrix} 3 & -3 & -6 & 5 \\ -3 & 2 & 5 & -4 \\ 2 & -6 & -4 & 7 \\ -3 & 0 & 5 & -2 \end{pmatrix}$$

From

```
>> E= eig(A)

E =

   -1.0000 + 0.0000i
   -1.0000 - 0.0000i
   -1.0000
    2.0000
```

we conjecture that -1 is an eigenvalue of multiplicity 3. If the Symbolic Toolbox is available, this is confirmed by

```
>> eigensys(A)

ans =

[ 2]
[-1]
[-1]
[-1]
```

We will use `rref` to compute the eigenvectors. First it will be convenient to introduce some shorthand.

```
>> I=eye(4);
>> r=-1;
>> rref(A-r*I)

ans =

     1     0     0     2
     0     1     0    -1
     0     0     1     1
     0     0     0     0
```

The fourth column is a free column. Setting $v_4 = -1$, and solving for the rest, we see that there is only one independent eigenvector

```
v1=[2 -1 1 -1];
```

Corresponding to $\mathbf{v}^1$ we have the solution

$$\mathbf{x}^1(t) = e^{-t}\mathbf{v}^1 = e^{-t}\begin{pmatrix} 2 \\ -1 \\ 1 \\ -1 \end{pmatrix}.$$

Notice that $\mathbf{x}^1(0) = \mathbf{v}^1$.

166

Next we use `rref` to compute the nullspace of $(A - (-1)I)^2$.

```
>> rref((A-r*I)^2)

ans =

     1     0    -2     0
     0     1     0    -1
     0     0     0     0
     0     0     0     0
```

Now there are two free variables. Since we have already used the fourth column, this time let's set this variable equal to 0, and set $v_3 = 1$. Then we get

```
>> v2 = [2 0 1 0]';
>> (A-r*I)*v2

ans =

     2
    -1
     1
    -1
```

and we recognize this answer as v1. Of course, this means that $(A - (-1)I)^2 v^2 = 0$. The solution corresponding to $\mathbf{v}^2$ is

$$
\begin{aligned}
\mathbf{x}^2(t) &= e^{tA}\mathbf{v}^2 \\
&= e^{-t}e^{t(A-(-1)I)}\mathbf{v}^2 \\
&= e^{-t}\left(\mathbf{v}^2 + t\,(A - (-1)I)\mathbf{v}^2\right) \\
&= e^{-t}\left(\mathbf{v}^2 + t\,\mathbf{v}^1\right) \\
&= e^{-t}\begin{pmatrix} 2+2t \\ -t \\ 1+t \\ -t \end{pmatrix}
\end{aligned}
$$

We still only have two solutions for the eigenvalue -1, which has algebraic multiplicity 3. To find the last one we look at the nullspace of $(A - (-1)I)^3$.

```
>> rref((A-r*I)^3)

ans =

    0    1    0   -1
    0    0    0    0
    0    0    0    0
    0    0    0    0
```

We need to choose a vector in the nullspace of this matrix, which is not a linear combination of $\mathbf{v}^1$ and $\mathbf{v}^2$. Since v_1 is now a free variable, we can set $v_1 = 1$, and we get

```
>> v3=[1 0 0 0]';
```

We have to calculate

```
>> (A-r*I)*v3

ans =

    4
   -3
    2
   -3
```

You might recognize this as $3\mathbf{v}^1 - \mathbf{v}^2$. We must also compute

```
>> (A-r*I)^2*v3

ans =

   -2
    1
   -1
    1
```

Again we notice that this answer is $-\mathbf{v}^1$.

168

We know that $(\mathbf{A} - (-1)\mathbf{I})^3\mathbf{v}^3 = \mathbf{0}$. Hence, by equation (11), our third solution is

$$\begin{aligned}
\mathbf{x}^3(t) &= e^{t\mathbf{A}}\mathbf{v}^3 \\
&= e^{-t}e^{t(\mathbf{A}-(-1)\mathbf{I})}\mathbf{v}^3 \\
&= e^{-t}\left(\mathbf{v}^3 + t\,(\mathbf{A} - (-1)\mathbf{I})\mathbf{v}^3 + \frac{t^2}{2}(\mathbf{A} - (-1)\mathbf{I})^2\mathbf{v}^3\right) \\
&= e^{-t}\left(\begin{pmatrix}1\\0\\0\\0\end{pmatrix} + t\begin{pmatrix}4\\-3\\2\\-3\end{pmatrix} + \frac{t^2}{2}\begin{pmatrix}-2\\1\\-1\\1\end{pmatrix}\right) \\
&= e^{-t}\begin{pmatrix}1 + 4t - t^2 \\ -3t + t^2/2 \\ 2t - t^2/2 \\ -3t + t^2/2\end{pmatrix}
\end{aligned}$$

Notice that $\mathbf{x}^3(0) = \mathbf{v}^3$.

Finally we compute the fourth solution using the eigenvalue 2. Since we know this eigenvalue has multiplicity 1, we can set

```
>> v4=null(A-2*I)

v4 =

    -0.3780
     0.3780
     0.3780
     0.7559
```

By examination we see that we can get integer entries by dividing by the first entry.

```
>> v4=v4/v4(1)

v4 =

     1.0000
    -1.0000
    -1.0000
    -2.0000
```

Thus our fourth solution is

$$\mathbf{x}^4(t) = e^{2t}\mathbf{v}^4.$$

To solve the initial value problem with

$$\mathbf{x}(0) = \begin{pmatrix} 2 \\ -1 \\ 0 \\ -1 \end{pmatrix},$$

We have to find the linear combination $\mathbf{x} = a_1\mathbf{x}^1 + a_2\mathbf{x}^2 + a_3\mathbf{x}^3 + a_3\mathbf{x}^4$, where the coefficients satisfy

$$\mathbf{V}\mathbf{a} = \begin{pmatrix} 2 \\ -1 \\ 0 \\ -1 \end{pmatrix},$$

and the matrix $\mathbf{V}$ contains the column vectors which are the initial values of the basic solutions. Thus

```
>> V=[v1 v2 v3 v4];
>> V\[2 -1 0 -1]'

ans =

    1
   -1
    2
    0
```

The solution is $\mathbf{x} = \mathbf{x}^1 - \mathbf{x}^2 + 2\mathbf{x}^3$.

Exercises

1. For each of the following matrices find the eigenvalues and eigenvectors. You may use any method you wish. For each eigenvalue find the algebraic and the geometric multiplicity.

a) $\begin{pmatrix} -6 & 0 \\ 7 & 1 \end{pmatrix}$ b) $\begin{pmatrix} 0 & 1 \\ 1 & 0 \end{pmatrix}$ c) $\begin{pmatrix} 0 & 1 \\ -1 & 0 \end{pmatrix}$ d) $\begin{pmatrix} 2 & -1 \\ 1 & 0 \end{pmatrix}$

e) $\begin{pmatrix} -2 & 2 & -1 \\ -6 & 5 & -2 \\ 4 & -2 & 3 \end{pmatrix}$ f) $\begin{pmatrix} 6 & -6 & 0 \\ -1 & 11 & -17 \\ 2 & 2 & -5 \end{pmatrix}$ g) $\begin{pmatrix} 3 & -2 & 1 & -2 \\ -2 & 1 & 1 & 2 \\ 2 & -4 & 3 & 0 \\ 7 & -5 & 1 & -5 \end{pmatrix}$

2. For each of the matrices $\mathbf{A}$ in exercise 1 find a basis of solutions to the system $\mathbf{x}' = \mathbf{A}\mathbf{x}$.

170

3. For each of the matrices $\mathbf{A}$ in exercise 1 parts a) through d) find the solution to the system $\mathbf{x}' = \mathbf{A}\mathbf{x}$ which satisfies the initial condition $\mathbf{x}(0) = \begin{pmatrix} 1 \\ 0 \end{pmatrix}$.

4. For the matrix in exercise 1 e) find the solution to the equation $\mathbf{x}' = \mathbf{A}\mathbf{x}$ which satisfies the initial condition $\mathbf{x}(0) = \begin{pmatrix} 0 \\ 2 \\ 1 \end{pmatrix}$.

5. For the matrix in exercise 1 f) find the solution to the equation $\mathbf{x}' = \mathbf{A}\mathbf{x}$ which satisfies the initial condition $\mathbf{x}(0) = \begin{pmatrix} 1 \\ 3 \\ 1 \end{pmatrix}$.

6. For the matrix
$$\mathbf{A} = \begin{pmatrix} 8 & -7 & 2 \\ -6 & 9 & -3 \\ -2 & 7 & -4 \end{pmatrix},$$

find the solution to the equation $\mathbf{x}' = \mathbf{A}\mathbf{x}$ which satisfies the initial condition $\mathbf{x}(0) = \begin{pmatrix} 1 \\ -2 \\ 1 \end{pmatrix}$.

7. For the matrix
$$\mathbf{A} = \begin{pmatrix} -6 & 6 & 3 \\ -10 & -9 & 6 \\ 3 & 0 & 8 \end{pmatrix},$$

find the solution to the equation $\mathbf{x}' = \mathbf{A}\mathbf{x}$ which satisfies the initial condition $\mathbf{x}(0) = \begin{pmatrix} 0 \\ -3 \\ -1 \end{pmatrix}$.

10. Phase Plane Analysis

A planar, autonomous system is a system of differential equations of the form

$$x' = f(x, y),$$
$$y' = g(x, y). \tag{1}$$

Thus we have two equations and two unknown functions. We will frequently use vector notation, and with $\mathbf{x} = \begin{pmatrix} x \\ y \end{pmatrix}$, and $\mathbf{f} = \begin{pmatrix} f \\ g \end{pmatrix}$, the two equations can be written as one vector equation

$$\mathbf{x}' = \mathbf{f}(\mathbf{x}). \tag{2}$$

The only way this differs from the general system of two equations in two unknowns is that the right-hand sides do not depend explicitly on the independent variable t. For systems of this sort, analysis of the *phase plane* can be very illuminating. The phase plane is simply (x, y) space, and we are interested in the parametric plots of the solution curves of the system.

For a general planar system of the form $\mathbf{x}' = \mathbf{f}(t, \mathbf{x})$, a solution is a function of the variable t, with values in $\mathbf{R}^2$. We can consider this function as the parametrization of a curve in the plane. The fact that the function is a solution of the differential equation means that at every point $(t, \mathbf{x})$, the curve must have $\mathbf{f}(t, \mathbf{x})$ as a tangent vector. For a fixed value of t we can imagine the vector $\mathbf{f}(t, \mathbf{x})$ attached to the point $\mathbf{x}$, representing the collection of all possible tangent vectors to solution curves for that specific value of t. Unfortunately the vector field changes as t changes, so this rather difficult visualization is not too useful. However, if the system is autonomous, then the vector field is not changing as t changes. Therefore, for an autonomous system the same vector field represents all possible tangent vectors to solution curves for all values of t. If a solution curve is plotted parametrically, at each point the vector field must be tangent to the curve.

There is a MATLAB function, `pplane`, that makes this visualization easy.* This chapter is a description of all of the features of `pplane`. Some of these features can be used immediately, while others require more advanced knowledge of ordinary differential equations. The student who is just beginning to learn about the subject may find that much of what is discussed here is a mystery. The way to read this chapter is to focus on what is needed at any one time and to refer back to it as your need and knowledge increases. If a term is used which is not understood, refer to your textbook for an explanation.

Actually, the functionality of `pplane` is very similar to that of `dfield`, so if you are familiar with that function, you will have no trouble with `pplane`.

---

* To see if `pplane` is installed properly on your computer enter `help pplane`. If it is not installed see the Preface for instructions on how to obtain it.

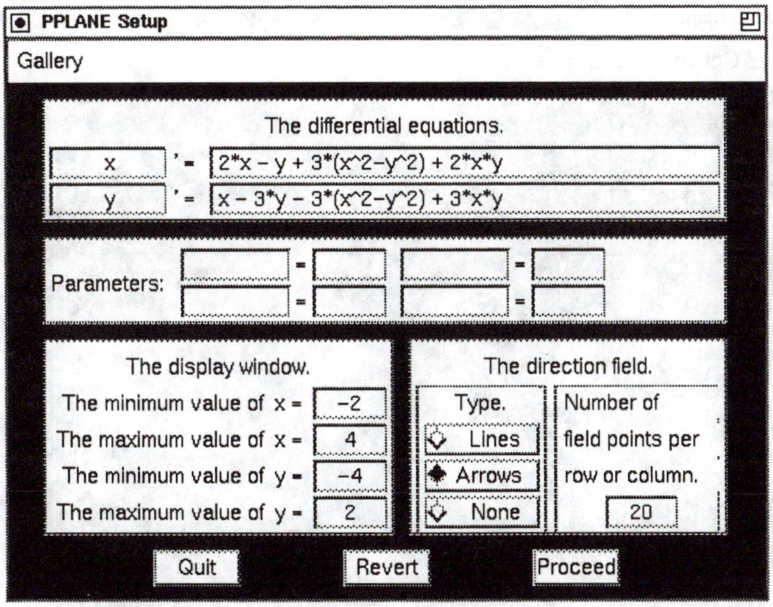

Figure 10.1. The setup window for `pplane`.

Starting `pplane`

To see `pplane` in action enter `pplane` at the MATLAB prompt. After a short wait, a new window will appear with the label **PPLANE Setup** . Figure 10.1 shows how this window looks on a UNIX machine. The appearance will differ slightly depending on your computer, but the functionality will be the same on all machines.

You will notice that there is a rather complicated autonomous system already entered in the upper part of the **PPLANE Setup** window. There is a middle section for entering parameters, although there are none entered at the moment. There is another section for describing a "display window", and yet another for defining what kind of field is required. At the very bottom there are three buttons. There will also be a new menu labeled **Gallery** . Its location will depend on your computer. On a UNIX machine it appears in the upper left-hand corner of the **PPLANE Setup** window, as shown in Figure 10.2. On a Macintosh or in PC-Windows it appears on the menu bar when **PPLANE Setup** is the active window. This menu contains a list of autonomous systems, and to enter one into the **PPLANE Setup** window, it is only necessary to select it from the menu.

We will describe the use of this window in detail later, but for now leave everything unchanged and click the button labeled | **Proceed** | . After a few seconds another window will open, this one labeled **PPLANE Display** . An example of this window is shown in Figure 10.2.

173

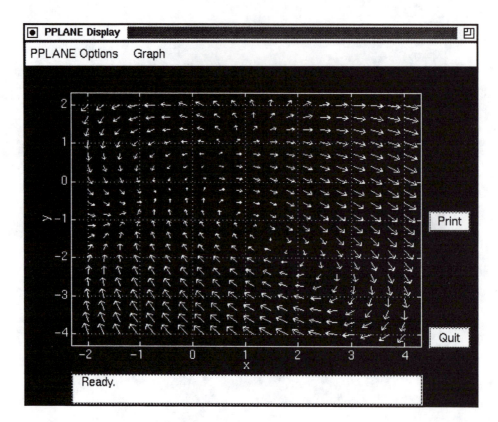

Figure 10.2. The display window for `pplane`.

The most prominent feature of the **PPLANE Display** window is a rectangle labeled with the variable x on the bottom, and the variable y on the left. The dimensions of this rectangle are slightly larger than the rectangle specified in the **PPLANE Setup** window. Inside this rectangle the **PPLANE Display** window shows the vector field for the equation which is defined in the **PPLANE Setup** window. At each point (x, y) of a grid of points, there is drawn a small arrow. The direction of the vector is the same as that of $\mathbf{f}(x, y)$, and the length varies with the magnitude of $\mathbf{f}(x, y)$. This vector must be tangent to any solution curve through (x, y).

There are two buttons on the **PPLANE Display** window, with the labels

Quit
 and
Print
 . There are two menus labeled **PPLANE Options** and **Graph** . Finally, below the vector field there is a message window through which `pplane` will communicate with the user. At this time it should contain the single word "Ready," indicating that it is ready to follow orders.

To compute and plot a solution curve from an initial point, move the mouse to that point, and click the mouse button. The solution will be computed and plotted, first in the direction in which the independent variable is increasing, and then in the opposite direction.

After computing and plotting a couple of solutions, the display will look something like Figure 10.3.

174

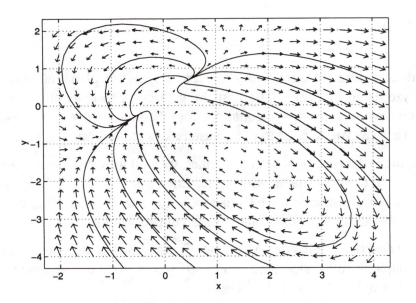

Figure 10.3. Several solutions to the equation.

Changing the differential system — using the PPLANE Setup window

We will illustrate the use of `pplane` by using it to do a phase plane analysis of the motion of a pendulum. The differential equation for the motion of a pendulum is

$$mL\theta'' = -mg\sin(\theta) - c\theta',$$

where θ is the angular displacement of the pendulum from the vertical, L is the length of the pendulum arm, g is the acceleration due to gravity, and c is the damping constant. If we choose a convenient measure of time by setting $s = \sqrt{g/L}\, t$, the equation becomes

$$\frac{d^2\theta}{ds^2} = -\sin(\theta) - a\frac{d\theta}{ds},$$

where

$$a = \frac{c}{m\sqrt{gL}}$$

is again called the damping constant.

We want to write this as a first order system, so we introduce the variables

$$x = \theta$$

$$y = \frac{d\theta}{ds}.$$

175

Then we have

$$x' = y$$
$$y' = -\sin(x) - ay,$$

where the prime indicates differentiation with respect to s. This is a planar autonomous system which we can analyze using pplane. The first thing to do is to enter this system into the **PPLANE Setup** window by entering the equations into the appropriate boxes, essentially the way they are printed above. We do need to use an asterisk ($*$) to indicate the product.

Since the damping constant a is not yet a number, we will have to assign it a value in one of the parameter boxes. It does not matter which one. Enter a in the left hand side and the value in the right hand side. For the time being, let's use the value $a = 0$. This value can be changed later to a positive number to see the phase plane for a damped pendulum.

Next we have to describe the display rectangle. Since $x = \theta$ represents an angle, and we will want to plot a couple of full periods, we enter $-10 \le x \le 10$. For the dimensions in y we have to experiment. For now $-4 \le y \le 4$ will do.

Finally, we decide which kind of direction field we want and how many field points we want displayed. These are the same options that were available in dfield. Let's keep the default values.

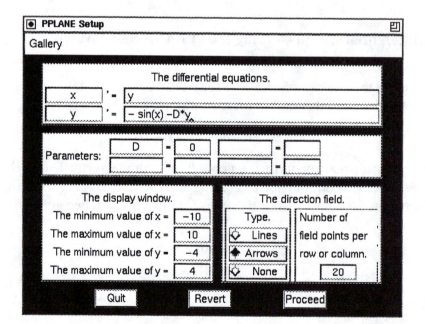

Figure 10.4. The **PPLANE Setup** window for the pendulum equation.

The completed **PPLANE Setup** window for the pendulum equation is shown in Figure 10.4. Now when we click on the ⬚ **Proceed** ⬚ button, we get the vector field for the

176

pendulum on the rectangle chosen. (Although entering the information about the pendulum equation is good practise, we could have have accomplished the same thing by selecting "pendulum" from the **Gallery** menu.)

At the very bottom of the **PPLANE Setup** window there are three buttons. We have already seen that the $\boxed{\textbf{Proceed}}$ button transfers the information in the **PPLANE Setup** window to the **PPLANE Display** window, and starts the computation of the direction field. Clicking the $\boxed{\textbf{Revert}}$ button will return all of the settings in the **PPLANE Setup** window to what they were when you began to make changes. This is useful, for example, when you have made a number of changes and you decide that it would be easiest to start over. The $\boxed{\textbf{Quit}}$ button should be used whenever you want to stop using `pplane`. It will close all related windows, and keep MATLAB running.

Plotting solution curves

Assuming that the **PPLANE Setup** window contains the information indicated in Figure 10.4, click the $\boxed{\textbf{Proceed}}$ button. After the **PPLANE Display** windows opens, click the mouse button at a point near $(x, y) = (0, 1)$. The computer begins to plot a solution curve. A closed orbit is plotted, giving the phase plane depiction of the standard motion of a pendulum.

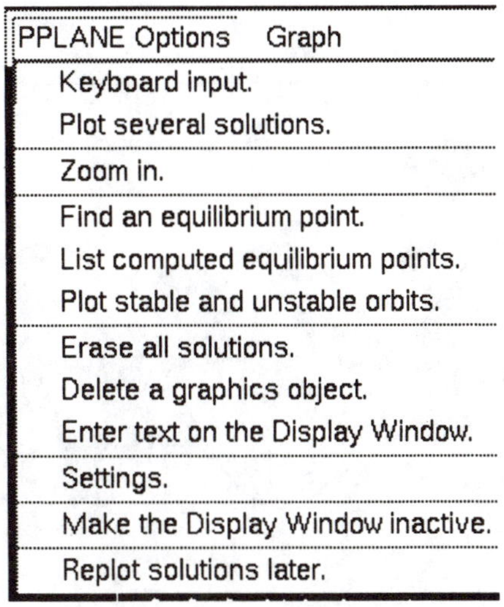

PPLANE Options   Graph

Keyboard input.

Plot several solutions.

Zoom in.

Find an equilibrium point.

List computed equilibrium points.

Plot stable and unstable orbits.

Erase all solutions.

Delete a graphics object.

Enter text on the Display Window.

Settings.

Make the Display Window inactive.

Replot solutions later.

Figure 10.5. The options menu.

Since we are interested here in explaining the operation of `pplane`, we will leave the interpretation of the solution curve to the reader. To further whet your curiosity, plot the curves beginning at $(0, 1.95)$, $(0, 2)$, and $(0, 2.05)$. It is difficult to achieve great precision by choosing

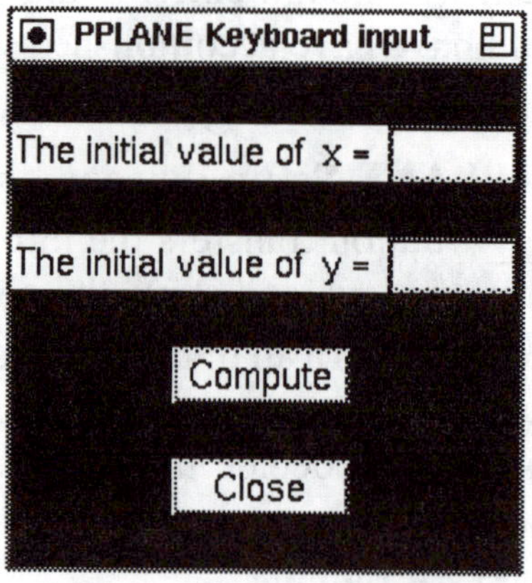

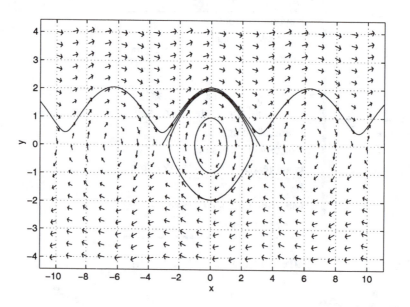

Figure 10.6. The Keyboard input window.

the initial point with the mouse. To allow greater precision, there is an option to enter initial data using the keyboard to be found in the **PPLANE Options** menu (See Figure 10.5).

If you select this option, another window will open labeled **PPLANE Keyboard input** (see Figure 10.6). Using this window very accurate initial conditions can be entered. Clicking on the ⟨ **Compute** ⟩ button will start the computation with the initial conditions entered. Enter the data for the three points.

Figure 10.7. Several solution curves for the pendulum.

After plotting these solutions, the Display Window should look like Figure 10.7.

You may not have noticed it, but while we were looking at the **PPLANE Display** window, pplane has been making comments on the Command Window. These comments describe features of the solution that pplane has noticed. Here is a list of the messages that can appear on the Command Window, and what they mean.

```
A nearly closed orbit was detected.
```

This means that a solution curve has circled around and come back close enough to a previously computed portion of the orbit that the computer was able to detect it. This could be a closed orbit, representing a periodic solution, or it could mean that the orbit is approaching a limit cycle (see Example 3 below for an example of a limit cycle). pplane sometimes returns this message in error when the solution is approaching an equilibrium point in a very broad spiral. We will see some examples of this.

```
The orbit ends in a possible equilibrium point near (-3.1, 0.0388).
```

Sometimes solution curves end (as $t \to \pm\infty$) at an equilibrium point. If pplane detects this happening it stops the computation and presents the above message. The coordinates it gives are not completely accurate, but the existence of an equilibrium point and its precise location can be verified with adequate accuracy by using the option to search for an equilibrium point (see below).

```
Maximum number (1300) of iterations reached.
```

To keep the computer from continuing to compute for too long a period on a very difficult orbit, the computation is stopped after 1300 steps, and the user is so informed.

```
A step size smaller than the minimum required.
```

The computer is not allowed to use a step size which is smaller than a minimum size (10^{-10}). If a smaller step size is required by the solution algorithm, it might indicate that the function $\mathbf{f}(\mathbf{x})$ has a discontinuity near the point where the computation stopped.

Choosing **Plot several solutions** allows you to select as many initial points with the mouse as you like, ending by hitting the **Return** key. All of the selected solution curves will be plotted.

Other plots of the solutions

The phase plane provides one view of a solution to an autonomous system. Often other views are also useful. pplane provides an easy way to display graphs of already computed solutions against the independent variable.

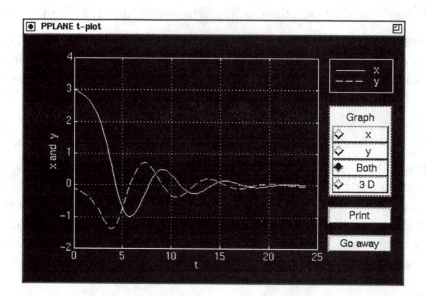

Figure 10.8. The **PPLANE t-plot** window.

Under the **Graph** menu on the **PPLANE Display** window, the user is presented with four options. Either of the solution variables can be plotted individually against the independent variable (which for convenience is assumed to be *t*), both of the variables can be plotted against the independent variable in a 2D plot, or both of the variables can be plotted against the independent variable in a 3D plot. After selecting one of these choices, the user is prompted to select a solution curve with the mouse. After this is done, a new window is opened with the name **PPLANE t-plot**, and in this window is the requested graph. An example is shown in Figure 10.8.

The user can choose between four views of the solution by the clicking one of the radio buttons with the mouse.

It is frequently the case that the default rectangle in which the graph is presented is not optimal. This can easily be changed using the `axis` command in the Command Window. For the 2D graphs the format is `axis([tmin, tmax, xmin, xmax])`, where `tmin` and `tmax` define the range of the independent variable, and `xmin` and `xmax` define that of the dependent variable(s). For the 3D graph the format is `axis([xmin, xmax, ymin, ymax, tmin, tmax])`.

Replotting solutions

The user will have noticed that after `pplane` has plotted a solution step-by-step, the entire display window is replotted. This is necessary because, while the step-by-step solution is present on the screen, it is not a part of the picture internally to the program. The solution must be

180

replotted before the display window is printed or copied, or the solution will be missing from the result.

Nevertheless, the constant replotting takes time. For that reason, immediate replotting can be disenabled, leaving the timing of replotting in the hands of the user. If the **Replot solutions later** option is chosen from the **PPLANE Options** menu (see Figure 10.5), subsequent solutions will not be replotted immediately. After the first such solution, a new button, labeled **Replot** will appear on the **PPLANE Display** window. Clicking this button will replot all solutions that need it. Notice that the **Replot** button is visible only if there is data that needs replotting. For example, after it is clicked, it disappears. There are other options that cause all of the solutions to be replotted, so the **Replot** button will sometimes disappear unexpectedly.

Immediate replotting can be reenabled by choosing the same option from the **PPLANE Options** menu. It is now labeled **Replot solutions immediately**.

Printing, quitting, and using clipboards

The easiest way to print the display window is to click the **Print** button. This will replot the solutions that need it, and issue the print command.

By now you realize that the display window can be printed by entering `print` at the MATLAB prompt, and in the Macintosh and the PC-Windows versions of MATLAB, the display window can be printed using methods that are standard to those operating systems. Before doing so you should ensure that all solutions have been replotted by clicking the **Replot** button, if it is visible. It is easier to click the **Print** button, which will replot the solutions and issue the print command. The same caution applies if you want to use the MATLAB option of saving the display window to a postscript file. After clicking the **Replot** button this can be accomplished from the Command Window in the usual way.

Sometimes MATLAB is very slow in arranging for a graphics window to be printed. While this is being done, strange things may happen to the window, such as the buttons disappearing. None of this is worrisome, unless you are really in a hurry. There is nothing you can do about it, so this is a time to be patient. Always wait until the word "Ready" appears in the message window before you try to do anything else with MATLAB. If you do not wait the results are unpredictable.

In the Macintosh and PC-Windows version of MATLAB, the contents of the display window can be copied into a clipboard, and from there into other documents in the standard ways. It is imperative that the **Replot** button be clicked before copying to a clipboard, or solutions curves may be missing from the copy.

When you want to quit `pplane`, the best way is to use the **Quit** buttons found on the **PPLANE Setup** and on the **PPLANE Display** windows. Either of these will close all of the `pplane` windows in an orderly manner, and it will delete the temporary files that `pplane` creates in order to do its business.

If on occasion you do not quit `pplane` before quitting MATLAB, temporary files can accumulate. These will have names like `pptp8765.m`. It is safe to delete these files, as long as you are not using `pplane` at the same time.

Equilibrium points

An *equilibrium point* is a point $\mathbf{x}_0 = (x_0, y_0)$ where $\mathbf{f}(\mathbf{x}_0) = 0$. If this is true, then the constant curve $\mathbf{x}(t) \equiv \mathbf{x}_0$ is a solution. This corresponds to a stationary point in the motion described by the differential system. For example, in the case of the pendulum, when the bob is hanging motionless below the point of suspension, we have a perfectly valid solution in which there is no motion. In the coordinates we are using this corresponds to $x = \theta = 2k\pi$, and $y = \theta' = 0$, where k is any integer.

The pendulum has another set of equilibrium points. If the bob is balanced directly above the suspension point, and there is no velocity, the bob should stay right there. Of course, this is a very unstable situation — the slightest disturbance will set the bob going. This corresponds to $x = \theta = (2k + 1)\pi$, and $y = \theta' = 0$.

An equilibrium point $\mathbf{x}_0$ is said to be *stable*, if every solution curve which starts close enough to $\mathbf{x}_0$ remains close to $\mathbf{x}_0$ as $t \to \infty$. It is said to be *asymptotically stable* if every solution curve which starts close enough to $\mathbf{x}_0$ approaches $\mathbf{x}_0$ as $t \to \infty$. An asymptotically stable equilibrium point is called a *sink*. If an equilibrium point is not stable, it is said to be *unstable*. A particular type of unstable point is a *source,* which has the feature that every solution curve which starts close enough to $\mathbf{x}_0$ approaches $\mathbf{x}_0$ as $t \to -\infty$.

Whether an equilibrium point is stable or not can **sometimes** be determined by looking at eigenvalues of the Jacobian. If the equation is $\mathbf{x}' = \mathbf{f}(\mathbf{x})$, where

$$\mathbf{f}(\mathbf{x}) = \begin{pmatrix} f(x, y) \\ g(x, y) \end{pmatrix},$$

then the *Jacobian* is

$$\mathbf{Jf} = \begin{pmatrix} \frac{\partial f}{\partial x} & \frac{\partial f}{\partial y} \\ \frac{\partial g}{\partial x} & \frac{\partial g}{\partial y} \end{pmatrix}.$$

If both eigenvalues of $\mathbf{Jf}$ evaluated at an equilibrium point have negative real part, then the equilibrium point is asymptotically stable (i.e., a sink). If at least one of the eigenvalues of $\mathbf{Jf}$ has a positive real part, then the critical point is unstable. If both eigenvalues of $\mathbf{Jf}$ evaluated at an equilibrium point have positive real part, then the equilibrium point is a source. It is possible to obtain other information about the equilibrium point from the Jacobian, as we will explain below.

A search for an equilibrium point can be initiated by selecting **Find an equilibrium point** from the **PPLANE Options** menu. You will be prompted to choose an approximate location of an equilibrium point. In the example of the pendulum, click the mouse near $(0, 0)$ or $(3, 0)$. Before very long a small circle will appear at the equilibrium point.

At the same time a new window opens with the label **PPLANE Equilibrium point data**. This window gives a variety of information about the equilibrium point. First it tells us something about the type of the equilibrium point. Then the Jacobian matrix of **f** at this point is displayed. Finally the eigenvalues and the associated eigenvectors of this matrix are computed and presented.

It is not necessary to be extremely accurate in choosing the point with the mouse. If `pplane` cannot find an equilibrium point close enough to the chosen point it will inform you in the message window.

`pplane` will recognize the types of many equilibrium points, but it is very conservative in its approach, and will often admit that it is unsure. Since `pplane` examines an equilibrium point numerically, and because numerical algorithms always make small truncation errors, such a conservative approach is required. In particular, `pplane` will only report information that is generic, i.e. information that will not change if the data on which it is based is changed by a small amount. Here is a complete list of the messages that can appear describing the equilibrium point, followed by an indication of what the message means.

`There is a saddle point at ...`

The eigenvalues are real and have different signs.

`There is a nodal sink at ...`

The eigenvalues are real, distinct, and negative. In some books this is called an *improper node,* a *stable improper node,* or a *stable node.*

`There is a nodal source at ...`

The eigenvalues are real, distinct, and positive. In some books this is called an *improper node,* an *unstable improper node,* or an *unstable node.*

`There is a spiral sink at ...`

The eigenvalues are complex conjugate and the real parts are negative. This is sometimes called a *spiral point*, a *stable spiral point,* or a *stable focus.*

`There is a spiral source at ...`

The eigenvalues are complex conjugate and the real parts are positive. This is sometimes called a *spiral point*, an *unstable spiral point,* or an *unstable focus.*

```
There is a spiral equilibrium point at ...
```

```
Its specific type has not been determined.
```

The eigenvalues are complex conjugate, with non-zero imaginary part. However, the real part is too close to 0 to decide whether this is a source, a sink, or a center. In particular any true center (i.e. conjugate, purely imaginary eigenvalues) will be included in this category. A center is also called a *focus* in some books.

```
There is a source at ...
```

```
Its specific type has not been determined.
```

The eigenvalues have positive real part, but the eigenvalues are too close together to decide the exact type of the source. This could be a spiral source, a nodal source, or a degenerate source for which the Jacobian has equal eigenvalues. An equilibrium point of this type is unstable, but not much more can be said.

```
There is a sink at ...
```

```
Its specific type has not been determined.
```

The eigenvalues have negative real part, but the the eigenvalues are too close together to decide the exact type of the sink. This could be a spiral sink, a nodal sink, or a degenerate sink for which the Jacobian has equal eigenvalues. An equilibrium point of this type is asymptotically stable, but not much more can be said.

```
There is an equilibrium point at ...
```

```
Its specific type has not been determined
```

At least one of the eigenvalues is so close to 0 that it is not possible to decide anything about the type of the equilibrium point. Closer analysis is required to describe the behaviour of the solutions.

It can be seen that pplane will label an equilibrium point only when it is sure. Of course, more precise information about an equilibrium point can sometimes be garnered by looking at the Jacobian matrix. The Jacobian as provided by pplane is a numerical approximation to the actual matrix of derivatives. Numerical approximations to derivatives are frequently inaccurate, so, especially in cases were there is doubt, the actual Jacobian should be computed and analyzed. The power of computers is still not sufficient to eliminate the need to do algebra.

However, looking at the Jacobian may not suffice, since this amounts to looking at the linearization of the system, and not the actual system. There are systems for which the linearization does not tell the whole story. Consequently, in the cases when pplane cannot decide the precise type, the ultimate test of the type is what the solution curves actually do. Even then it is entirely possible that visualization will not provide the answer, and a rigorous analysis will be necessary. See Example 2 for a case where the linear analysis does not provide the definitive answer. Other examples are given in the exercises.

As pplane computes equilibrium points, it remembers them. To see the current list, choose **List computed equilibrium points** from the **PPLANE Options** menu. The list will be printed to the Command Window. If you want a hard copy, you can obtain one by using the `diary` command.

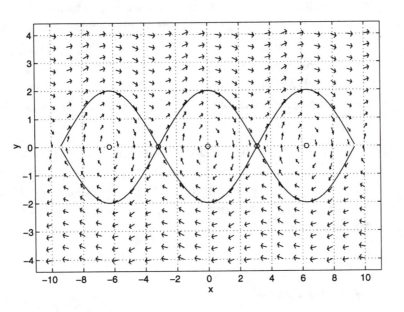

Figure 10.9. Separatrices and equilibrium points.

Stable and unstable orbits

In a neighborhood of any saddle point, there are four special orbits. The two *stable* orbits approach the saddle point as $t \to +\infty$. The two *unstable* orbits approach the saddle point as $t \to -\infty$. Together these orbits are called *separatrices*. In the case of the pendulum, the stable orbits correspond to the bob nearing the point directly above the suspension point, but never quite reaching it in any finite time. This can happen from either side of the top. The unstable orbits correspond to the bob being just nudged off the top at $t = -\infty$.

pplane will calculate and plot separatrices. We already know that the pendulum has a saddle point at $(\pi, 0)$, so first choose the **Plot stable and unstable orbits** from the **PPLANE Options** menu. Then click the mouse near $(\pi, 0)$. The computer will begin to compute and plot the orbits. Meanwhile you will be kept informed of what is happening by messages in the information window. After plotting the separatrices from $(-\pi, 0)$ and a couple of other equilibrium points, the result will be similar to Figure 10.9.

Changing the size and appearance of the display window

Some people prefer to use a direction field in the **PPLANE Display** window, instead of the vector field. Such a change can be made in the **PPLANE Setup** window. It is simply a matter of clicking on the button of your choice. At the same time the number of field points displayed can be changed. The default is 20 points in each row and in each column. Any integer between 5 and 40 will be acceptable. Similarly a different display rectangle can be chosen in the **PPLANE Setup** window. When all of the new options are selected, click the $\boxed{\textbf{Proceed}}$ button. However, you must be careful not to change any of the information about the differential equation or the parameters. If such a change is made, even if it is changed back to the original, pplane assumes that you want to start over, and it will erase all of the previously computed orbits and equilibrium points.

Choosing the **Zoom in** option from the **PPLANE Options** menu provides a convenient way to focus your attention on a portion of the current display rectangle. After making this selection, you can choose a smaller display rectangle by clicking on two opposite corners of the new rectangle with the mouse. There is no zoom out option, since this can be accomplished using the **PPLANE Setup** window.

Personalizing the display window

Sometimes when you are preparing a display window for printing, you plot a solution curve you wish were not there. There are ways to correct for this. There are two choices on the **PPLANE Options** menu which allow you to erase items from the display. The **Erase all solutions** option is self-explanatory.

The second one, **Delete a graphics object**, is much more flexible. It will allow you to delete any solution curve. It is only necessary to choose this option, and then to select the solution curve by clicking the mouse anywhere on the curve. It is best to choose a place on the solution curve which is as far away as possible from other curves or even line field elements. It is very easy to confuse pplane, in which case the wrong curve, or even a direction field element might be deleted instead of the curve you want. It is possible using this option to delete any graphics object on the display window, including text.

There are three text elements which are part of the display window. These are the title at the top, the xlabel at the bottom, and the ylabel at the left. The last two are given default values by pplane using the information entered into the **PPLANE Setup** window. All three can be changed at any time.

It is also possible to add text at arbitrary points in the display window. To do this choose **Enter text on the Display Window** from the **PPLANE Options** menu. You will be prompted to enter the desired text in the Command Window. Now when the mouse cursor is in the display window it changes from the normal arrow to a cross. Position the cross at the point where you want the lower left part of the text to appear, click the mouse button, and the text will appear. You should choose a position for the text which is outside of the vector field, since text is hard to read on top of the vector field. If the placement of your text does not please you, you can remove the text using the **Delete a graphics object** option.

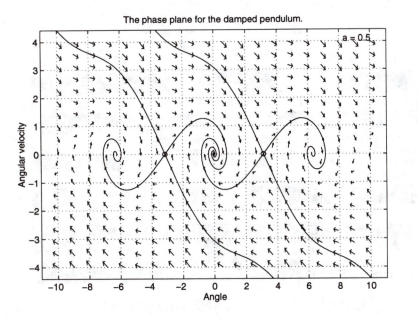

Figure 10.10. Separatrices for the damped pendulum.

As an example illustrating all of this, suppose you have plotted several solutions of the damped pendulum equation, and you decide that you want to change the labels. Here are the commands to enter in the Command Window to change the title and the axis labels to reflect your desires.

```
>> xlabel('Angle')
>> ylabel('Angular velocity')
>> title('The phase plane for the damped pendulum.')
```

Now suppose you decide that you want to indicate the value of the damping constant somewhere on the figure. Choose the **Enter text on the Display Window** option from the menu, and when you are prompted in the Command Window, enter something like

```
>> Enter the text here. >  a = 0.5
```

The results will be similar to Figure 10.10.

Computational speed, accuracy, and kinky plots

The **PPLANE Setup** window gives the user options that affect the speed and accuracy of computation, and the appearance of the solution curves. All of these involve tradeoffs. The **PPLANE Settings** window can be opened by choosing **Settings** from the **PPLANE Options** menu. It is shown in Figure 10.11.

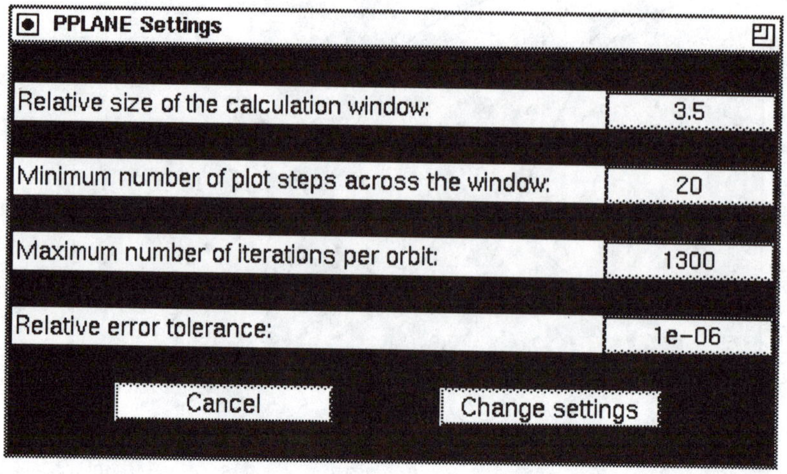

Figure 10.11. The PPLANE settings window.

The design of `pplane` includes the definition of two windows. We have already discussed the display window. The calculation window is 3.5 times as large as the display window in each dimension. The computation of a solution does not stop when the solution curve leaves the display window, but rather when it leaves the calculation window. The larger calculation window allows for reentrant solutions, i.e., solution curves which leave the display window at the top, the bottom, or one of the sides, and later return to it. It also allows for zooming to a larger display window without having incomplete solution curves.

The first item in the **PPLANE Settings** window controls the relative size of the calculation window. Any value bigger than or equal to 1 is valid. The smaller this number, the faster `pplane` will compute solutions, but it is more likely that reentrant solutions will be lost. It is not unreasonable to set this value to 1 if you have a slower computer, and if you are not going to be zooming to a larger display window.

`pplane` computes solutions using a numerical method that is closely related to the routine `ode45` which we discussed in Chapter 6. It finds approximate solution points, and connects them with straight lines segments. If the points are too far apart, the curve might look kinky. The distance in the independent variable between two points is called the *step size*. Within `pplane` the step size is controlled by two parameters which can be set in the **PPLANE Settings** window. The first is the **Minimum number of plot steps across the window.** If we call this

188

number N, then the step size is at most $1/N$ times the width of the display window. The second is called the **Relative error tolerance**. If we denote this number by T, then roughly speaking, the step size is chosen to ensure that the estimated error being made at any step is less than T times the absolute value of the solution.

The larger the value of N, and the smaller that of T, the more accurate the solution. On the other hand, increasing N or decreasing T increases the number of steps, and as a result the computation is slower. N can be any non-negative integer. When N is small, the step size can be large and the solution curves may have kinks, even if the tolerance T is small. As you can see, the trade off between computation speed and the appearance of the solution curves is quite complicated.

The default settings should be sufficient in most cases. If you are using a very fast computer, you should consider setting the tolerance T to be equal to 10^{-5} or 10^{-6}, i.e. `1e-5` or `1e-6` in MATLAB's parlance. If your computer is not so fast, and the speed with which solutions are being computed bothers you, set $N = 0$, and perhaps even $T = 0.001$. You can then decrease T as you feel necessary to increase your accuracy, and to make the solution curves smoother.

The final setting in the **PPLANE Settings** window is the **Maximum number of iterations per solution**. The default value is 1300. There are very few solution curves that will require this many steps. If `pplane` uses this many steps, it will stop computing and so inform you by a message in the Command Window, but it will have plotted the solution up to the point where the computation stopped. If this happens, and you want more information about the solution that was being computed, you can increase this number and try again.

A rarely needed option

It may happen that you want to click the mouse while its cursor is in the display window, without plotting a solution curve. If this is the case, choose **Make the Display Window inactive** from the **PPLANE Options** menu. When you want to plot more solution curves choose this option again — it is now labeled **Make the Display Window active.** It is not necessary to do this before using any of the `pplane` options.

Using MATLAB while `pplane` is open

This is certainly possible. There is an avoidable problem that can arise if you try to plot something while `pplane` is open. You may discover that you are plotting to one of the `pplane` windows, with your graph appearing right on top of what is already there. To avoid this, simply enter `figure` at the MATLAB prompt before you plot for the first time. A new Figure Window will open with nothing in it, and with a title like **Figure No. nn**, where nn is a number on the order of 3 or 4.

Future plotting commands will be directed to this window, as long as you do not click the mouse while the cursor is in another window. If, by chance, you should select another window in this way, you can change back to the right Figure Window, either by clicking the mouse in the correct Figure Window, or by entering `figure(nn)` at the MATLAB prompt.

Example 1

By examining the equilibrium points and the stable/unstable orbits, a rather complete understanding of a planar autonomous system can be attained. Perhaps the reader has already noticed this with the pendulum example. Here we will present a couple of other examples.

Consider the system

$$x' = y\cos(x^2 - y^2 + 1),$$
$$y' = -x\cos(x^2 + y^2).$$

We will look at this system in the rectangle $-2 \le x \le 2$, $-2 \le y \le 2$. After calling `pplane` with this data we see a rather complicated vector field. However, by systematically looking for equilibrium points where the direction field changes direction rapidly, and then plotting the stable/unstable solution curves through the saddle points we find, we get to the point where we can almost completely understand the system using the result.

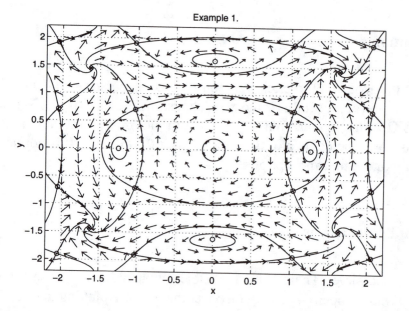

Figure 10.12. Final analysis of Example 1.

190

We can see a list of all of the equilibrium points that have been calculated by selecting that option from the **PPLANE Options** menu. There are 25 of them in this rectangle.

About the only question left is a precise description of the motion around the five equilibrium points that are labeled `Spiral equilibrium point`. To discover this, we plot a solution near each of these points. We discover that the orbits are closed, and, consequently, each of these points is actually a center. Finally, having the sense that there are too many arrows, we change the number to 18 in the **PPLANE Setup** window.

The final result is Figure 10.12. Now, given any initial point in the display window we can use the fact that no two orbits can meet each other (except at equilibrium points for $t = \pm\infty$) to predict, at least qualitatively, what the solution curve through that point will look like.

Example 2

Consider the system
$$x' = y\cos(x - y + 1),$$
$$y' = -x\cos(x + y).$$

We will be particularly interested in the behavior of this system near the origin, so choose the rectangle $-2 \le x \le 2$, $-2 \le y \le 2$.

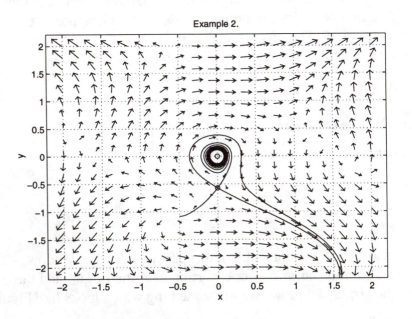

Figure 10.13. A weak spiral source at the origin.

191

First examine the equilibrium point at the origin. `pplane` says it is a spiral equilibrium point. Direct computation shows that the Jacobian matrix is

$$\begin{pmatrix} 0 & \cos(1) \\ -1 & 0 \end{pmatrix}.$$

The eigenvalues are $\pm i\sqrt{\cos(1)}$, so the linear analysis says this is a center.

Next notice that there is a saddle point at approximately $(0, -0.5)$. After plotting the stable/unstable orbits we reach Figure 10.13. Notice that one of the stable orbits spirals in towards the origin very slowly, and stops with the message on the Command Window saying `A nearly closed orbit was detected`. In this case `pplane` is making a mistake. If this orbit is continued, it will continue to spiral into the origin. Remembering that this is a stable orbit for the saddle point, we conclude that the orbit starts at the origin at $t = -\infty$, spirals outward as t increases, and approaches the saddle point as $t \to \infty$. Consequently, although the linear analysis indicates that the origin is a center, in fact it is a very weak spiral source.

There are other equilibrium points in the display window, and finding them will be left as an exercise.

Example 3

Until now all of the solution curves we have encountered have either been closed curves, or they have approached ∞ or an equilibrium point, as $t \to \pm\infty$. It is appropriate to ask whether these are the only possibilities. The answer is that there are others. The most important is illustrated by an equation which was devised by Lord Rayleigh to model the motion of the reed in a clarinet:

$$x'' = ax' - b(x')^3 - kx,$$

where a, b, and k are positive constants.

Setting $y = x'$, this becomes the system

$$x' = y$$
$$y' = ay - by^3 - kx$$

With $a = 5$, $b = 4$, and $k = 5$ we enter this into `pplane` with the window $-2 \le x_1 \le 2$, $-2 \le x_2 \le 2$. Plot two solution curves — one starting near the origin, and the other starting near the boundary of the display window. We get something which looks like Figure 10.14.

For both solutions (and in fact for all solutions) as t increases the solution approaches a closed curve. This closed curve is what is called a *limit cycle*. The limit cycle itself is a solution to the system. It corresponds to a periodic solution of the second order equation, and it can be interpreted as the steady tone that a clarinet produces.

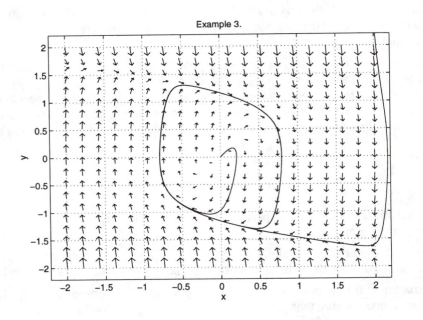

Figure 10.14. Rayleigh's model of the clarinet.

Exercises

1. Consider the differential system

$$\mathbf{x}' = \mathbf{A}\mathbf{x},$$

where **A** is one of the matrices

i) $\begin{pmatrix} 1 & 1 \\ -18 & 10 \end{pmatrix}$ ii) $\begin{pmatrix} 6 & -1 \\ 0 & -3 \end{pmatrix}$

iii) $\begin{pmatrix} -1 & 0 \\ 3 & -3 \end{pmatrix}$ iv) $\begin{pmatrix} 6 & 1 \\ -18 & 0 \end{pmatrix}$

v) $\begin{pmatrix} 9 & -1 \\ 9 & 3 \end{pmatrix}$ vi) $\begin{pmatrix} 9 & 1 \\ 9 & 3 \end{pmatrix}$

vii) $\begin{pmatrix} 7 & 4 \\ -10 & -5 \end{pmatrix}$ viii) $\begin{pmatrix} -1 & 0 \\ 0 & -1 \end{pmatrix}$

ix) $\begin{pmatrix} 6 & -4 \\ 18 & -6 \end{pmatrix}$ x) $\begin{pmatrix} 2 & -2 \\ 4 & -4 \end{pmatrix}$

a) In each of these cases find the type of the equilibrium point at the origin. Do this without the aid of `pplane`. You may, of course, check your answer using `pplane`.

b) For each of these cases use `pplane` to plot several solution curves — enough to fully illustrate the behavior of the system. You will find it convenient to use the "linear system" choice from

193

the **Gallery** menu. If the eigenvalues are real, use the "Keyboard input" option to include solutions starting at ± 10 times the eigenvectors. If the equilibrium point is a saddle point, compute and plot the separatrices.

2. For the linear systems i) through iv) in problem 1, consider the effect of a non-linear perturbation which leaves the Jacobian at $(0,0)$ unchanged. The theory predicts that, in (maybe very small) neighborhoods of equilibrium points where the eigenvalues of the Jacobian are non-zero and not equal, a non-linear system will act very much like the linear system associated with the Jacobian. In this exercise we will use `pplane` to verify this. To be precise, instead of i) consider the perturbed system

$$x' = x + y - xy,$$
$$y' = -18x + 10y - x^2 - y^2.$$

   a) Show that the Jacobian of the perturbed system at the origin is the same as that of the unperturbed system.

   b) Starting with the display rectangle $-5 \le x \le 5$, $-5 \le y \le 5$, zoom in to smaller squares centered at $(0,0)$ until the solution curves look like those of the linear system in i).

   Repeat the experiment for the systems in ii), iii), and iv) using the same or other quadratic or cubic terms as perturbations. Be sure that the perturbation vanishes to second order at the origin.

3. In contrast to the previous problem, consider the system

$$x' = y + ax^3$$
$$y' = -x$$

   for the three values 0, 10 and -10 of the parameter a.

   a) Show that all three systems have the same Jacobian matrix at the origin.

   b) Use `pplane` to find evidence that will enable you to make a conjecture as to the stability of the equilibrium point at $(0,0)$ in each of the three cases.

   c) Consider the function $h(x,y) = x^2 + y^2$. In each of the three cases, restrict h to a solution curve and differentiate the result. Can you use the result to verify the conjecture you made in part b)?

   d) Does the Jacobian predict the behavior of the non-linear systems in this case?

4. Consider the system

$$x' = ax + y$$
$$y' = -x - 2y$$

   Describe what happens to the equilibrium points as a varies among the five choices -0.1, -0.01, 0, 0.01, and 0.1. Provide plots of each with enough solution curves to illustrate the behavior of the system. If the eigenvalues are real, use the **Keyboard input** option to include solutions starting at ± 1 times the eigenvectors.

5. For each of the solution curves for the pendulum indicated in Figure 10.7 describe the motion of the pendulum. If you have trouble understanding the motion from the phase plane diagram, use the option in the **Graph** menu to plot the angular displacement versus time.

6. Consider the damped pendulum when the damping constant $a = 0.5$, and when $a = 2.5$.

   a) Use `pplane` to examine the behavior of the system in each of these cases.

   b) You will notice that the type of the equilibrium point at the origin is different for $a = 0.5$ and $a = 2.5$. Analyze the Jacobian matrix and find out at which value of a the type changes.

7. This is another example of how a system can change as a parameter is varied. Consider the system

$$x' = ax + y - x(x^2 + y^2)$$
$$y' = -x + ay - y(x^2 + y^2)$$

for $a = 0$, and $a = \pm 0.1$. Use the display rectangle $-1 \le x \le 1$, and $-1 \le y \le 1$, and plot enough solutions in each case to describe behavior of the system. Describe what happens as the parameter a varies from negative values to positive values. (This is an example of a *Hopf bifurcation*).

8. MATLAB's animation feature provides an attractive way to view the Hopf bifurcation of the previous problem. Here we will explain how do do this without going into the details. First it is necessary to create a derivative M-file for the Hopf system. It should have the parameter a as a global variable. Here is what it should look like:

```
function xpr = hopf(t,x)

global A

xpr(1) = A*x(1)+x(2)-x(1)*(x(1)^2+x(2)^2);
xpr(2) = -x(1)+A*x(2)-x(2)*(x(1)^2+x(2)^2);
```

Next create the following M-file, which will create the data for the movie, and call it hopfmov.m

```
N = 9;              % Number of frames.
K = 4;              % Number of solution curves.
T = 40;             % The smallest final time.

dela = 0.2/(N-1);        % Step size in A.

global A

axis([-1,1,-1,1]);
hold on
M = moviein(N);
for k=1:N
        A = -0.1+(k-1)*dela
        cla
        for j = 1:K
                init = [cos(2*j*pi/K),sin(2*j*pi/K)];
                tf = T +(.1-abs(A))*2600;
                [t,x] = ode45('hopf',0,tf,init);   % Old version.
                plot(x(:,1),x(:,2));drawnow
        end
        M(:,k) = getframe;
end
```

If your computer has a lot of memory, you can increase the number of frames to 17. After executing hopfmov, the matrix M will contain the data needed for the movie. Executing movie(M,10) will show it on your computer screen.

195

9. This problem involves the default system in `pplane`, i.e. the planar system that appears when `pplane` is first started.

   a) Find and plot all interesting features, i.e., equilibrium points, separatrices, limit cycles, and any other features that appeal to you. Make a list of such to turn in with the plot.

   b) Find the linearization of the system at $(0, 0)$. Enter that into `pplane` and plot a few orbits.

   c) Go back to the default system. According to theory, the linearization should approximate the original system in a small enough neighborhood of $(0, 0)$. Use the **Zoom in** option to find a small enough rectangle containing $(0, 0)$, where the behavior of the system is closely approximated by the linear system you analyzed in part b).

   d) Redo parts b) and c) for the other equilibrium points of the default system.

10. The default system in `pplane` exhibits two bifurcations involving the saddle point at the origin and the equilibrium point near $(3/2, -3/2)$. To be precise we are looking at the system

$$x' = 2x - y + 3(x^2 - y^2) + 2xy$$
$$y' = x - 3y - 3(x^2 - y^2) + Axy$$

as the parameter A varies.

   a) Show that between $A = 2.5$ and $A = 3$, the equilibrium point changes from a sink to a source and spawns a limit cycle, i.e., a Hopf bifurcation occurs.

   b) Show that between $A = 3$ and $A = 3.2$ the limit cycle disappears. At some point in this transition, the limit cycle becomes a homoclinic orbit for the saddle point at the origin. A *homoclinic orbit* is one that originates as an unstable orbit for saddle point at $t = -\infty$, and then becomes a stable orbit for the same saddle point as $t \to +\infty$. It is extremely difficult to find the value of A for which this occurs, so do not try too hard. The process illustrated here is called a *homoclinic bifurcation*.

   c) The default system has four quadratic terms. Show that, if any of the coefficients of these terms is altered up or down by as little as 0.5, a bifurcation takes place. Show that there is a Hopf bifurcation in one direction, and a homoclinic bifurcation in the other.

If $\mathbf{x}_0$ is a sink for the system $\mathbf{x}' = \mathbf{f}(\mathbf{x})$, then the *basin of attraction* for $\mathbf{x}_0$ is the set of points $\mathbf{y} = \begin{pmatrix} y_1 \\ y_2 \end{pmatrix}$ which have the property that the solution curve through $\mathbf{y}$ approaches $\mathbf{x}_0$ as $t \to +\infty$.

11. Consider the damped pendulum with damping constant $a = 0.5$. Find the basin of attraction for $\begin{pmatrix} 0 \\ 0 \end{pmatrix}$. You should indicate this region on a printed output of `pplane`. **Hint:** It will be extremely helpful to plot the stable and unstable orbits from a couple of saddle points.

12. The system of differential equations:

$$x' = \mu x - y - x^3$$
$$y' = x$$

is called the *van der Pol system*. It arises in the study of non-linear semiconductor circuits, where y represents a voltage and x the current. It is in the **Gallery** menu.

   a) Find the equilibrium points for the system. Use `pplane` only to check your computations.

b) For various values of μ in the range $0 < \mu < 5$, find the equilibrium points, and find the type of each, i.e, is it a nodal sink, a saddle point, ...? You should find that there are at least two cases depending on the value of μ. Don't worry too much about non-generic cases. Use `pplane` only to check your computations.

c) Use `pplane` to illustrate the behavior of solutions to the system in each of the cases found in b). Plot enough solutions to illustrate the phenomena you discover. Be sure to start some orbits very close to $\begin{pmatrix} 0 \\ 0 \end{pmatrix}$, and some near the edge of the display window. Put arrows on the solution curves (by hand after you have printed them out) to indicate the direction of the motion. (The display window $(-5, 5, -5, 5)$ will allow you to see the interesting phenomena.)

d) For $\mu = 1$ plot the solutions to the system with initial conditions $x(0) = 0$, and $y(0) = 0.2$. Plot both components of the solution versus t. Describe what happens to the solution curves as $t \to \infty$.

13. *Duffing's equation* is
$$mx'' + cx' + kx + lx^3 = F(t).$$

When $k > 0$ this equation models a vibrating spring, which could be soft ($l < 0$) or hard ($l > 0$) (see Problem 5.6). When $k < 0$ the equation arises as a model of the motion of a long thin elastic steel beam that has its top end embedded in a pulsating frame (the $F(t)$ term), and its lower end hanging just above two magnets which are slightly displaced from what would be the equilibrium position. We will be looking at the unforced case (i.e. $F(t) = 0$), with $m = 1$.

a) This is the case of a hard spring with $k = 16$, and $l = 4$. Use `pplane` to plot the phase planes of some solutions with the damping constant $c = 0, 1$, and 4. In particular, find all equilibrium points.

b) Do the same for the soft spring with $k = 16$ and $l = -4$. Now there will be a pair of saddle points. Find them and plot the stable/unstable orbits.

c) Now consider the case when $k = -1$, and $l = 1$. For each of the cases $c = 0$, $c = 0.2$, and $c = 1$, use `pplane` to analyze the system. In particular find all equilibrium points and determine their types. Plot stable/unstable orbits where appropriate, and plot typical orbits.

d) With $c = 0.2$ in part c), there are two sinks. Determine the basins of attraction of each of these sinks. Indicate these regions on a print out of the phase plane.

14. Suppose that we have a solution to the planar autonomous system

$$x' = f(x, y),$$
$$y' = g(x, y).$$

At a point $\begin{pmatrix} x \\ y \end{pmatrix}$ on the solution curve where $f(x, y) \neq 0$, we have

$$\frac{dx}{dt} = x' = f(x, y) \neq 0,$$

hence by the inverse function theorem, the function $x(t)$ has an inverse $t(x)$, and

$$\frac{dt}{dx} = 1 / \frac{dx}{dt} = 1/f(x, y).$$

If we define $y(x) = y(t(x))$, then by the chain rule,

$$\frac{dy}{dx} = \frac{dy}{dt}\frac{dt}{dx} = \frac{g(x, y)}{f(x, y)}.$$

Thus the solution curves in the phase plane to the system

$$\begin{aligned} x' &= f(x, y) \\ y' &= g(x, y) \end{aligned},$$

are the same as the solution curves to the single first-order equation

$$\frac{dy}{dx} = \frac{g(x, y)}{f(x, y)},$$

at least near points where the denominator $f(x, y) \neq 0$.

The exercise is to plot solutions to the equation $\frac{dy}{dx} = -x/y$ using `dfield`, and solutions to the corresponding system using `pplane` and compare the results. You should use the same display windows and compare the line fields. Then find solutions with the same initial points in each case (something like $x = 0$, $y = 1$). Do not be taken aback if the Command Window keeps telling you that you are trying to divide by zero. It's true, you are trying to divide by zero, but you can't hurt anything. Then find and compare the analytic solutions to each.

15. Repeat the previous exercise for the equation

$$xy\frac{dy}{dx} = y - 1.$$

Use the window $0 \leq x \leq 4$, $-3 \leq y \leq 1$, and initial points $\begin{pmatrix} 1 \\ -1 \end{pmatrix}$, $\begin{pmatrix} 2 \\ -1 \end{pmatrix}$, and $\begin{pmatrix} 1 \\ 1/2 \end{pmatrix}$. Again you can find the exact solution.

16. Complete the analysis of the system in Example 2 by finding as many of the equilibrium points as you can in the display window and plotting stable/unstable orbits where appropriate. There are 10 equilibrium points in the rectangle.

17. A wide variety of phenomena can occur when an equilibrium point is completely degenerate, i.e., when the Jacobian is the zero matrix. We will look at just one. Consider the system

$$\begin{aligned} x' &= xy \\ y' &= x^2 - y^2. \end{aligned}$$

a) Show that the Jacobian at the origin is the zero matrix.

b) Plot the solutions through the six points $(0, \pm 1)$, and $(\pm\sqrt{2}, \pm 1)$. Plot additional solutions of your choice.

c) Compare what you see with the behavior of solutions near a saddle point.

Index

Items that appear in typewriter font, such as `dfield`, refer to MATLAB commands or variables. Those in bold-face, such as **Keyboard input**, are the names of windows, or menus, or menu items. The names of buttons are boxed, such as ☐ Compute ☐ .

soft spring, 100, 197.

solution curve, 11, 174.

solution, approximate, 71.

solutions, reentrant, 19, 188.

`solve`, 59 – 61, 104 – 119, 149.

solving linear equations, 126 – 144.

source, 182, 184.

  degenerate, 184.

  nodal, 183.

  spiral, 183.

span, 141.

speed, 18.

spiral equilibrium point, 184, 192.

spiral point, 183.

spiral sink, 183.

spiral source, 183, 192.

spring constant, 99.

`sqrt`, 3, 6, 101.

stable, 23, 102, 182, 185.

  focus, 183.

  improper node, 183.

  node, 183.

  orbit, 185.

  spiral point, 183.

stable/unstable orbits, 190, 192.

steady state solution, 100, 109.

step size, 19, 70, 95, 179, 188.

stiff equations, 97.

stopping MATLAB, 95.

Student Edition of MATLAB, x, 50.

submatrix, 124.

`subs`, 63, 106, 107, 108, 110, 111, 112, 113.

subscripting options, 89.

`sym`, 58, 111.

`symadd`, 53, 54.

`symbolic`, 50.

symbolic algebra, 50.

symbolic algebra program, 104.

symbolic calculations, 1.

symbolic computing, *ix, xi.*

symbolic expression, 50.

symbolic routine, 153.

Symbolic Toolbox, 1, 50, 104 – 119, 149, 151, 154, 159, 166.

symbolic vector, 151.

`symdiv`, 53, 54.

`symmul`, 53, 54.

`symop`, 54, 64, 110.

`symsub`, 53, 54.

`symvar`, 55.

`symvar` rule, 55 – 56.

system of homogeneous equations, 140.

systems of equations, determined, 135.

systems of equations, overdetermined, 135.

systems of equations, underdetermined, 135.

systems of first order equations, 86.

systems of linear equations, 121.

temporary files, 15, 182.

text, 17.

text string, 17.

thunderhead, 90.

`title`, 17, 21, 41, 42, 43, 75, 78, 86, 186.

tolerance, 19, 85, 93, 95.

transient, 100.

transpose, 127.

truncation errors, 183.

`type`, 47.

type of equilibrium point, 183.

underdetermined systems of equations, 135.

uniqueness, 26.

UNIX, 1, 10, 173.

unstable, 24, 103, 182, 185.

  focus, 183.

  improper node, 183.

  node, 183.

  orbit, 185.

  spiral point, 183.

up arrow, 48.

van der Pol equation, 97.

van der Pol system, 97, 196.

variables, local, 46.

variation of parameters, 114 – 116.

vector, 29.

  column, 30.

  row, 30.

vector field, 16.

vector of right hand sides, 35.

vector of unknowns, 35.

vector space, 30.

`vectorize`, 118.